MODELING GOD'S WILLS

J.G. Lenhart

WITH COMMENTARY BY

Dr. Joel Swokowski

Copyright © 2024 by John Lenhart

All rights reserved.

No part of this publication may be reproduced, distributed, or transmitted in any form or by any means, including photocopying, recording, or other electronic or mechanical methods, without the prior written permission of the publisher, except as permitted by U.S. copyright law. For permission requests, contact info@modelinggod.com.

Although the publisher and the author have made every effort to ensure that the information in this book was correct at press time and while this publication is designed to provide accurate information in regard to the subject matter covered, the publisher and the author assume no responsibility for errors, inaccuracies, omissions, or any other inconsistencies herein and hereby disclaim any liability to any party for any loss, damage, or disruption caused by errors or omissions, whether such errors or omissions result from negligence, accident, or any other cause.

All Bible scriptures are taken from the King James Version unless noted otherwise

Commentary by Dr. Joel Swokowski

Original background art by Joe Bailen
Cover art by Rob Warning

Book design by Mark D'Antoni, eBook DesignWorks

ISBN: 979-8-9927804-3-7

www.modelinggod.com

Table of Contents

Preface 9
   Dr. Joel Swokowski's Commentary 19

Book 1 — God's Wills **24**
   PART ONE — How God Brings About His Will 25
      Chapter 1 — The Party 26
         Modeling God Review 27
         GPS Analogy 29
         Calvinism Comparison 32
         The Biggest System 34
         Summary 36
         Dr. Joel Swokowski's Commentary 36

      Chapter 2 — God's Plan 41
         Modeling God 46
         Dr. Joel Swokowski's Commentary 49

      Chapter 3 — God's Contrastive Process 52
         Eden 54
         Pre-Flood 54
         Post-Flood 55
         Abraham 56
         The Law 57
         Pentecost 59
         Millennium 61
         Dr. Joel Swokowski's Commentary 64

      Chapter 4 — Calvinism 68
         Predestination 70
         Dr. Joel Swokowski's Commentary 78

      Chapter 5 — Moses and Jeremiah 86
         Jeremiah 90

Summary 93
   Dr. Joel Swokowski's Commentary 94

Chapter 6 — Ezekiel and Hezekiah 96
   Hezekiah 101
   Dr. Joel Swokowski's Commentary 105

Chapter 7 — God's Will Explanation 109
   Summary 114
   Dr. Joel Swokowski's Commentary 116

PART TWO — Characteristics of God's Will 120
   Chapter 8 — Individual Will 121
      Individual Will 126
      A Personal God 128
      Qualifying For More 129
      Summary 131
      Dr. Joel Swokowski's Commentary 131

   Chapter 9 — Group Wills 133
      Moses 134
      Jeremiah 134
      Ezekiel 135
      Hezekiah 135
      Implication 135
      Universal Will 137
      Summary 138
      Dr. Joel Swokowski's Commentary 140

   Chapter 10 — Three Levels in God's Will 143
      Perfect Will 144
      Acceptable Will 146
      Good Will 148
      Out of God's Will 150
      Summary 151
      Dr. Joel Swokowski's Commentary 153

   Chapter 11 — Council Meetings 156
      Leader vs. Boss 156
      Wise Man 158

- Councils 159
- Husbanding 163
- Summary 165
- Dr. Joel Swokowski's Commentary 166

Chapter 12 — Dissolve 171
- Solomon 173
- Red Sea 176
- Jesus 178
- Summary 180
- Dr. Joel Swokowski's Commentary 181

Chapter 13 — God's Four Measures for Judgment 184
- Solomon 186
- Blessed Examples 191
- Summary 193
- Dr. Joel Swokowski's Commentary 194

Chapter 14 — Prophecy 198
- Jonah 200
- Simple vs. Complex 203
- Summary 206
- Dr. Joel Swokowski's Commentary 207

Chapter 15 — How God Brings About His Will 209
- God's Story 211
- Our Story 212
- God's Resolution 217
- Summary 220
- Dr. Joel Swokowski's Commentary 221

Afterword 225
- Dr. Joel Swokowski's Commentary 227

Book 2 — Determining God's Wills 229

PART ONE — Intimacy and Communication 230

Chapter 1 — Intimacy 231
- Relationship vs. Fellowship 235
- Fellowship With God 239
- Dr. Joel Swokowski's Commentary 243

Chapter 2 — Covenants 248
   Noah 249
   Abram 250
   Moses. 251
   Jesus 252
   Breaking vs. Ending a Covenant. 253
   Summary. 256
   Dr. Joel Swokowski's Commentary 257

Chapter 3 — Communication 260
   Communication Causes 264
   Bullies 266
   Summary. 269
   Dr. Joel Swokowski's Commentary 270

Chapter 4 — Repair 273
   Repair Process. 274
   Forgiveness. 275
   God's Ability to Move. 278
   Dr. Joel Swokowski's Commentary 282

Chapter 5 — Transformation 286
   Church And Marriage 289
   Summary. 295
   Dr. Joel Swokowski's Commentary 296

PART TWO — Roadblocks to God's Will 300
  Chapter 6 — Marriage. 301
   Traditional Marriage. 303
   Whose Will? 304
   Biblical Marriage 305
   God's Marriage Perspective. 307
   Marriage Doctrine. 310
   Dr. Joel Swokowski's Commentary 312

  Chapter 7 — Marriage Covenant 315
   Marriage. 317
   Effects Of Sexual Intercourse. 318
   Fornication. 322
   Summary. 324
   Dr. Joel Swokowski's Commentary 325

Chapter 8 — Divorce 330
   Biblical Divorce 332
   God's Perspective. 336
   Making Divorce Easy? 337
   Dr. Joel Swokowski's Commentary 339

Chapter 9 — Putting Away 343
   Repairing Marriage 347
   God's Approach 349
   Summary. 352
   Dr. Joel Swokowski's Commentary 353

Chapter 10 — Divorce vs. Putting Away 357
   New Testament 360
   Jesus and Pharisees 362
   Jesus' Identity 367
   Summary. 369
   Dr. Joel Swokowski's Commentary 370

Chapter 11 — Church 375
   Business or Ministry. 376
   Acts Church 379
   Church Leadership 383
   Conclusion. 384
   Dr. Joel Swokowski's Commentary 385

Chapter 12 — Church Leadership 393
   Apostle 394
   Prophet. 396
   Apostles and Prophets 398
   Evangelist 399
   Pastor. 400
   Teacher. 406
   Summary. 407
   Dr. Joel Swokowski's Commentary 408

Chapter 13 — Women. 417
   Help Meet 417
   Cleave 418
   Last Creation 419

    Sarai And Abram 420
    Leah and Rachel 421
    Proverbs 31. 421
    1611 KJV I Esdras. 424
    Dr. Joel Swokowski's Commentary 428

Chapter 14 — Men. 432
    Definition of Man 433
    The Flip 438
    Ruth. 440
    Proverbs 31. 442
    Summary. 444
    Dr. Joel Swokowski's Commentary 445

Chapter 15 — Bringing About God's Will 451
    Facilitating God's Will. 453
    The Answer. 465
    Summary. 468
    Dr. Joel Swokowski's Commentary 469

The Next Book. 476

Preface

Modeling God's Wills is the second edition of a three-part series meant to pick up from where the first edition, *Modeling God*, left off. *Modeling God*, like this edition, also consisted of two books. The first of the two books ("Modeling God and Salvation") identified four tools for determining a comprehensive worldview (way of looking at reality) that presents and proves the only possible explanation for a supreme being and salvation. The second book ("Christian Living, Modeling Life") applied this worldview to everyday interactions and presented the meaning of life.

The overall goal of the three-edition series is to present a model that explains God's eternal plan, how God operates in order to bring about His plan, and how the concepts people talk about when expressing their beliefs fit into this worldview.

In this second edition, the first of the two books ("God's Wills") shows how God brings about His eternal plan, which we are calling God's will. The second book ("Determining God's Will") shows how we can determine if we are in God's will to the point we can help God bring about His eternal plan.

The importance of this topic can best be appreciated by understanding systems thinking. God created systems. He didn't create the universe to stand alone and not affect anything else. He didn't create man to stand alone and not affect anything else. Everything God created was in relation to a system and how they relate to other systems. If we want to understand God's will, we need to understand systems.

Let's begin by understanding the definition of a system according to a pioneer and innovative expert: Dr. Russell Ackoff.

> **1. System: A whole made up of two or more essential parts.**
>
> I hope you can see that everything appears to be a system. For example, you are a system made up of skin, bones, organs, etc. A car is a system made up of thousands of parts. Let's look at the next point from Dr. Ackoff.
>
> **2. The defining property of the system can only be exhibited by the whole and it is not exhibited by the parts.**
>
> For instance, the defining property for a car is transportation; however, no part of the car can provide transportation. In fact, the engine can't even transport itself. The next point is immediately understood from the previous point.
>
> **3. When you disassemble the whole, it loses its defining property.**
>
> The moment you take a car apart, you cease to have a vehicle that can provide transportation.

Dr. Ackoff didn't invent the perspective that results from the first three points. In fact, we have a record of the Apostle Paul making the same three points in I Corinthians 12:12-26 when he spoke of members and the body in place of parts and the whole:

> *For as the body is one, and hath many members, and all the members of that one body, being many, are one body: so also is Christ. For by one Spirit are we all baptized into one body, whether we be Jews or Gentiles, whether we be bond or free; and have been all*

made to drink into one Spirit. For the body is not one member, but many. **(System: A whole made up of two or more essential parts.)**

If the foot shall say, Because I am not the hand, I am not of the body; is it therefore not of the body? And if the ear shall say, Because I am not the eye, I am not of the body; is it therefore not of the body? If the whole body were an eye, where were the hearing? If the whole were hearing, where were the smelling? But now hath God set the members every one of them in the body, as it hath pleased him. And if they were all one member, where were the body? But now are they many members, yet but one body. **(The defining property of the system can only be exhibited by the whole, and it is not exhibited by the parts.)**

And the eye cannot say unto the hand, I have no need of thee: nor again the head to the feet, I have no need of you. Nay, much more those members of the body, which seem to be more feeble, are necessary: And those members of the body, which we think to be less honourable, upon these we bestow more abundant honour; and our uncomely parts have more abundant comeliness. For our comely parts have no need: but God hath tempered the body together, having given more abundant honour to that part which lacked. That there should be no schism in the body; but that the members should have the same care one for another. And whether one member suffer, all the members suffer with it; or one member be honoured, all the members rejoice with it. **(When you disassemble the whole, it loses its defining property.)**

Paul saw the church as a system made up of Jesus, the Holy Spirit, and believers, with believers consisting of two parts: Jews and Gentiles. The defining property is exhibited by the whole and not the individual parts. Focusing only on the individual parts would hinder the whole from achieving its defining property in God's eternal plan. (We will cover the defining property of the church in the second book of this edition.)

The fourth and final point concerning systems thinking is:

4. The way to improve the system is to improve the interaction between the parts.

Notice, Jesus said this in Matthew 5:29-30:

> *And if thy right eye offend thee, pluck it out, and cast it from thee: for it is profitable for thee that one of thy members should perish, and not that thy whole body should be cast into hell. And if thy right hand offend thee, cut it off, and cast it from thee: for it is profitable for thee that one of thy members should perish, and not that thy whole body should be cast into hell.*

Systems thinking believes the profitability is not in the part. The profitability is in the interaction between parts. Likewise, Jesus saw you as a system and stated that if one of your parts is controlling you and sinning, its interaction with the other parts will cause the whole body to be unprofitable enough to go into hell. Notice, if you are unable to get control over the faulty part so it interacts in a righteous way where the whole body profits, then you ought to remove it.

Jesus is a systems thinker, shown by the amount of focus He put on our interactions with others. Whether it is giving to the poor, forgiving others, or having your cloak taken, Jesus is very focused on the interactions both between people with other people and between people with God the Father.

Modeling God showed salvation is dependent on our interactions with God, while reward is dependent on our interactions with other people. *Modeling God* showed that uniqueness is what made profitability possible because profitability only results from the interaction between two people who value things differently.

Dr. Ackoff would pose a scenario to accentuate the importance of interactions over the intrinsic ability of the individual. Imagine a group of engineers working together to determine the carmaker that made the best engine, transmission, brakes, etc. For example, Mercedes may make the best transmission, while Ford makes the best brakes.

Dr. Ackoff would then ask what would you have if you gathered together all of the best parts of a car. You wouldn't even have a car! Why? Because the parts wouldn't fit together.

While Dr. Ackoff's perspective was on making the interactions better, the Bible (including the two passages above) came from the perspective of destructive interactions. In order to illustrate the Bible's perspective, imagine a company made up of two employees (Ann and Beth).

The profitability is completely dependent on the interaction between Ann and Beth. Let's say the profit from Ann and Beth is only able to deliver 95% of what the company needs, so we need to hire another employee (Curt) who would easily be able to deliver the last 5% by himself. Unfortunately, it turns out the company is now only delivering 50% of what it needs to in order to survive. What happened?

How many interactions are there now that we added a new employee? There are seven. There is one interaction where all three interact with each other at the same time according to their own beliefs (Ann-Beth-Curt).

There are three one-on-one interactions (Ann-Beth, Ann-Curt, Beth-Curt). That's only four total. What are the other three interactions?

Have you ever seen this situation? You get along with one friend when it is just you and them. You get along with another friend when it is just you and them. However, when those two friends get together, you don't get along with either of them? When your two friends become like-minded, they become a system that is unique from each of them individually. So, there are also three interactions where two of the employees are like-minded (AnnBeth-Curt, AnnCurt-Beth, BethCurt-Ann).

How many interactions is the new employee involved in? Six. If these additional interactions are a -45%, the profitability of the company drops from 95% to 50%. This is the reason behind the truth of the saying, "One bad apple spoils the whole bunch."

Now, imagine a community of millions with one bad apple. How many interactions involve the bad apple? Hundreds of millions. Remember what Jesus said about the part that is a "bad apple" in the body. If God's eternal plan involves creating more over time with a body made up of Jesus, the Holy Spirit, and millions of people, He can't afford to have even one soul present who isn't interested in interacting with others in a profitable manner. Now we see Jesus' advice about the body spoke to God's perspective on His eternal plan.

Modeling God stated the people who wanted to be in God's eternal plan would focus on growing in taking more direction from God via grace, and confess and repent when they didn't. That means "bad apples" would try to do their own plan and refuse admitting they were wrong and/or avoid repairing the damage they did.

Before we close this preface, we need to contrast two approaches: analysis vs. synthesis.

The analysis process consists of three steps:

1. Take the whole apart
2. Improve/understand each part
3. Put the system back together

For example, when it comes to fixing a car, we follow analysis. We take the problem area apart, fix the malfunctioning part, and reassemble the area.

Notice, the process of analysis requires the system (the whole) to cease to exist during steps 1 and 2. More importantly, this process focuses on the tangible, which means when it is used as a first step, at best, it can only restore the system in the moment and delay the inevitable destruction resulting from an intangible cause. *Modeling God* called this approach the "Survival Philosophy."

Our brains naturally follow the analysis process as a first step according to four attributes resulting in categories:

- **Compression:** Identifying a subjective category so you can treat unique members the same.
- **Amplify:** Categories result in an amplification of differences between artificially created groups.
- **Discriminate:** Categories result in groups being valued preferentially, while justifying the devaluing of other groups.
- **Fixed mindset:** Categories hinder future creative approaches and revelatory perspectives.

The reality is this analytical, categorical process works great on inanimate objects. For example, while looking for a used headlight for our car to replace our faulty one, we are going to sort through a series of headlights according to brand, age, etc. However, when it comes to applying analysis to people according to skin color, we amplify the fixed mindset of stereotyping and discrimination. We become racist. For example, at what skin color does a person become "white"? Good thing there is another approach!

The synthesis process consists of three steps:

1. Look to the next bigger system.
2. Define the purpose of the next bigger system.
3. Define the (purpose of the) original system within and relative to the next bigger system.

For example, when it comes to a car being the original system:

1. We identify "transportation" as the next bigger system.
2. We understand the purpose of transportation as transporting something from one location to another.
3. We define the purpose of a car as a personal, autonomous, safe-against-the-weather way of transporting an individual.

Notice, the process of synthesis keeps the system intact while we look to the greater system in order to improve or revolutionize the original system. More importantly, this process focuses on the intangible, which means it can repair the system while also avoiding the inevitable destruction resulting from an intangible cause. *Modeling God* called this approach the "Life Philosophy."

The only way to fully understand a system is to move to the next greater system. For example, people tend to only be able to learn 25% of information when they are taught directly. This is why people are told to repeat something three times or apply it three times before they can be sure they learned the information.

What is the next bigger system from learning that includes learning as a part? Teaching, which is pretty easy to see because it is a cause of learning. It turns out, when people teach a subject, they learn four times faster. Dr. Ackoff tells the story of an elementary school class that spent half a school year teaching mathematics to a computer and when they were tested at the completion of the project, it was found the students had learned two full years of mathematics.

I didn't realize it at the time, but the first analogy in the preface of *Modeling God* demonstrated the power of synthesis vs. analysis! It began with imagining a wall that is 200 miles long and 50 miles high. The wall is made up of tiles that are hundreds of feet square. As you read the following excerpt from *Modeling God*, realize the wall is God's eternal plan:

Most people focus on becoming experts on the details of a belief system. They walk closer to the wall. These books are attempting to take you far enough away from the wall that you get the big picture. This first book will only show you the wall isn't entirely blue. As you make your way through the books, the concepts in the first two books should begin to present an image. No doubt you will notice some tiles are missing, however, not enough to take away from the overall image. I will leave the filling in of these tiles for subsequent books.

If one continued moving away from the wall, they would find the vast number of tiles overwhelming. It would appear the wall is nothing more than a random collection of colored tiles. At this point, the individual must make a hard decision—continue this uncomfortable journey away from the wall or return to the comfort of the crowd.

If they persevered and moved further from the wall, they would find the tiles make up a picture of a man standing on grass. It now appears the wall is a mural and is actually trying to portray an image.

The previous edition left the reader at that uncomfortable halfway point where they couldn't pick out the tiles nor the greater images. This edition will take you further from the wall so you can understand the images that represent God's will. I'm sure many people will not fully comprehend this preface or the first book of this edition during their first read-through as they will still be at the uncomfortable stage. However, I have found if the reader continues to choose a growth mindset and venture further from the wall, they will have an epiphany as an effect of being transformed.

This is not easy because it goes against how our brain naturally works. In fact, every time the reader finds themselves wanting more information as to how something specifically works, they are taking a step towards the wall. This is analysis.

Remember, this edition is intended to step you away from the wall to show how the entire picture connects: synthesis. This will be done by bringing all the images on the wall into focus since the images represent God's will. The reality is these individual images are made up of several parts (chapters) and each of the individual chapters could be an entire book. It is very understandable that you may want more information, especially if the chapter presents a concept that is completely new to you.

The other reason the reader may want more information is they don't understand what has been presented. My advice is to first make sure the answer to your question wasn't already presented previously in the chapter. In fact, sometimes the answer may be in the previous paragraph!

Otherwise, the answer is ahead of you, much like when you watch a movie, after all, a movie is a system. Do you stop watching a movie the moment you don't understand something? No, you embrace the tension that comes from waiting for more information. Likewise, this book presents a system, and the full answer requires all of the parts, which means, they are ahead of you.

Dr. Joel Swokowski's Commentary

Editions and Books

People can get confused with the introduction because the author sees *Modeling God* as the first edition of a three-part series, which consisted of two books: "Modeling God and Salvation" and "Christian Living, Modeling Life." The wall excerpt above came from *Modeling God*, so the reference to the first book referred to "Modeling God and Salvation." The third edition will be titled "Modeling God's Plan." Also, all scripture passages presented in this edition, unless otherwise noted, are from the King James Version of the Bible. All definitions for words from the Bible will come from Strong's Concordance with their reference number, unless otherwise noted.

God's Plan and God's Will

Although the terms "God's will" and "God's plan" are often used interchangeably, they are different and will be used in specific fashions throughout this edition. Here are the definitions of each term:

- God's plan: the result of what God will accomplish, both in the moment and His eternal desire. This is the wall in Lenhart's analogy.
- God's will: how God accomplishes His plan; not "what He does," but the methods, manners, guiding principles, or processes for accomplishing His plan, both in the moment and according to His eternal desire. These are the images on the wall in Lenhart's analogy (e.g., a man teaching a man to fish).

For example, my plan may be to get a quart of milk from the store. However, my will (method) would be that you get it for me, but I pay for it. You getting it for me and you paying for it would not be in my will, but it would still accomplish my plan.

Plan: having milk.
Will: you get it, I pay for it.

Or how about this example that many of you would be familiar with: my plan may be to sit on a warm beach drinking cocktails with umbrellas and fruit. However, my will (method) would be to retire with enough money to make that happen.

Plan: sit on a beach and relax.
Will: retire with enough money to accomplish my plan.

Fundamental Christianity

As we venture into this edition and see the complexity of God's will and God's plan, it's important to note the foundation of what has been set from *Modeling God*:

- We learned that God's Nature is always completely Righteous and always completely Just. No matter the complexity of where we go from here, that remains true.

- We learned that humanity's connection to God is by grace through faith. Christians are meant to grow in taking more direction from a God they can't see in the hopes of a benefit they've yet to receive. This remains true as we unpack the ultimate benefit: God's eternal plan.

- Doing bad things doesn't make you a bad person. We all do bad things. How we respond to our behavior is more important than the behavior itself. A Christian is someone who responds to their bad behavior with confession and repentance. This remains true from the day a person becomes born again and as that person grows in making disciples of all nations.

Fundamental Christianity is simple but not easy:

1. Do what God is telling you to do.
2. Confess and repent when you don't.

This doesn't change just because what we're learning and what God is calling us to do becomes more complex.

Systems Thinking: Analytical vs. Synthesis

We were presented systems thinking with two different approaches: analysis and synthesis. As you read this for the first time, this may seem foreign or new. I assure you, it will become more clear as you read the ensuing chapters. For now, I will try to keep it simple and remember the "wall" analogy that Lenhart taught in the first edition. As much as you may be tempted to take a step forward to get details about a specific part, try to equally consider that your issue is that you're not seeing the whole picture. Take a step back before taking a step forward.

Now, I also want to clear up one more thing: there is nothing wrong with analytical thinking, in and of itself. Synthesis and analysis are both necessary. What Lenhart repeatedly stated is that synthesis ought to be the first step, followed by analysis. I hope no one thought that analysis was wrong, after all, the third step of the synthesis process is analysis! (Likewise, the third step of the analysis process is synthesis.)

The problem is that our brains naturally work according to analysis, and when we work with anything that does not involve humans, we can seem to get away with analysis as a first step. However, believing that understanding something requires analysis is the concept that has deceived people for thousands of years.

This deception would mean that if analysis is the key to understanding, then complete analysis is the key to complete understanding. But what happens when you take everything apart to its fundamental elements? You end up with nothing! What Lenhart taught in the very first analogy of *Modeling God's* preface was that the only way to understand anything completely is to use synthesis as a first step.

Even when it comes to inanimate objects, if analysis is all we do, we end up breaking the system instead of repairing it. Taking apart an engine is great in order to find out what's wrong with it. But if I keep breaking down the engine and putting it back together (even if I do it right), I'm adding undue wear and tear to the engine that will actually limit its life in the long term.

The best objective, or goal, with analytical thinking is when we use our understanding of the individual parts to facilitate restoration in the system, now and in the future. Yet, it's synthesis that will lead to us understanding all systems, from the smallest to the greatest system.

Analysis, even at its most beneficial, will never result in us understanding God's will at the level God desires. This means analytical thinking ought to be done as a second step, with synthesis thinking being the first step. Whenever you get uncomfortable while reading this edition, I suggest you ask yourself what the first step you took in your thinking was. Was it analysis or synthesis?

This second edition continues the synthesis approach. The third edition (*Modeling God's Plan*) will finish this three-part series with the analysis approach by filling in the missing tiles Lenhart spoke about in *Modeling God*.

Book 1

God's Wills

PART ONE

How God Brings About His Will

INSIDE

Chapter 1: The Party 26

Chapter 2: God's Plan 41

Chapter 3: God's Contrastive Process 52

Chapter 4: Calvinism 68

Chapter 5: Moses and Jeremiah 86

Chapter 6: Ezekiel and Hezekiah 96

Chapter 7: God's Will Explanation 109

CHAPTER 1

The Party

THE ANALOGY GUIDING each of these three editions was presented in the following manner at the beginning of *Modeling God*, Chapter 1: The Invitation.

Imagine you are invited to a party where you will be able to participate in any sensual pleasure you desire for as long as you want.

It's an exclusive party, but the invitation I hand you simply says "Admit One" and has a place for you to fill in your name. As you look at the invitation, you notice it contains no other information. The first question you would probably ask is, "When is the party?"

You are relieved to find out the party was several months away and you had plenty of time to fit it into your schedule. Your next question would probably be, "Where is the party?"

To this, I might say, "Chicago."

As I turn to leave, you stop me so you can ask, "Where in Chicago?"

When I answer, "In a building with a sign in front. Why do you ask?" you may begin to show frustration.

"Why do I ask? I want to know how to find it, that's why!"

My answer would depend on where you are coming from. If you are a great distance away, I'd tell you to first take a plane. Then, I'd tell you to drive a car. As you try to interrupt, I say, "Let me be more helpful— when you drive the car, one pedal is the gas. That makes the car go. The other pedal is the brake. That makes the car stop. There is also a steering wheel. That helps you direct the car."

If you haven't given up, you may ask, "What am I supposed to do, drive every street until I find it? Could you please give me an address?" The reason you want an address is to know exactly where the party is so that you can find it intentionally. Another benefit of an address is that it allows you to measure your progress.

Up until you get the address, this process would be totally unacceptable for directions to the party. However, these are the same type of directions we get from people when we ask how to get closer to God. Think about it—the most important desire of your heart and you settle for directions that you'd consider ridiculous for something less important.

Modeling God Review

The refusal by "experts" to explain how to be intentional with our growth in God fueled *Modeling God*. We found the address was Right and Just because this is God's Nature. Basically, the party was located at the corner of Righteous Street and Justice Avenue and we can measure our progress towards God's party with whether we are growing and becoming more right and just.

When we incorporated salvation by grace through faith into the model, we realized salvation is actually the same process as improving our relationship with God and intentionally getting closer to the party. So, the party is not salvation. Salvation is part of the journey to the party. We called it "The Door." The party represents a relationship with God that is perfect (maximum profitability), and it appears to be something we will continue to work on once we are in His presence because the profitability grows infinitely.

Book 2 focused on answering the question, "What is the journey?" and on how we intentionally make and measure progress, because it looks like we need an objective way to measure progress. We saw that profitability is God's universal measure for determining whether people are allowed to continue the journey once they die. Our ability to make it to the party depends on continual improvement in our interactions with Him. We need to let Him continually direct our actions via grace and faith. It is a process, not a one-time event.

This explained why our success in getting to the party depends on understanding Jesus more, not less. Consequently, for the rest of the journey, we will measure everything using profitability to ensure our progress toward the party—that is, to make sure we are "good." God is not looking for us to be without flaw. He is looking to see if we are growing.

Another implication for this explanation is that all of us are going to eventually get the maximum profitability out of our ARE. (Remember, from *Modeling God*, your ARE is who God created you to BE specifically.) This is the party. How is it possible for all of us to achieve this at the same party?

We've seen that Paul compared each of us to a part of the body. When each part does its role perfectly by having God flow through them, then the whole body benefits. We are supposed to develop the unique ability we were given by God (our ARE) in order to do our job perfectly and fit in perfectly with others who are doing their job, which results in maximum profitability. This is the journey we continue after we physically die.

This is God's plan. This is what it is all about. This is why we are here. This is the meaning of life! Your ultimate goal is to find and operate in your ARE, which is God working through you. This is your calling. This is your purpose.

God's plan in eternity is the party, which will consist of beings in covenant with several other beings through their ARE. This will require the ultimate in intimacy and communication. Every being will know what they are valued for and how much they value every other unique individual. There won't be jealousy, because everyone will be growing in perfection in his or her uniqueness.

GPS Analogy

This current edition will look at some of the characteristics and details of how God's eternal plan will work. In keeping with the analogy, this edition is focused on some of the characteristics and details of our journey.

> *⁸ For by grace are ye saved through faith; and that not of yourselves: it is the gift of God:*
> *⁹ Not of works, lest any man should boast.*
> *¹⁰ For we are his workmanship, created in Christ Jesus unto good works, which God hath before ordained that we should walk in them.* (Ephesians 2:8-10)

Verses 8 and 9 are the foundation of salvation and were covered in depth in the first edition. Look closely at verse 10.

It appears God set out the good works He wanted you to progress in before you were born. In our analogy, God had a specific route He wanted you to take to the party! More specifically, it appears your car has a GPS system with a programmed route. The directions are being sent all the time, which is grace.

In keeping with our analogy, these directions are sent to our car and picked up by an antenna, which converts the invisible information into tangible actions that are communicated to us via the GPS. What is the antenna? Let's look at John 3.

> *³ Jesus answered and said unto him, Verily, verily, I say unto thee, Except one be born again, he cannot see the kingdom of God.*
> *⁴ Nicodemus saith unto him, How can a man be born when he is old? can he enter the second time into his mother's womb, and be born?*
> *⁵ Jesus answered, Verily, verily, I say unto thee, Except a man be born of water and the Spirit, he cannot enter into the kingdom of God.*
> *⁶ That which is born of the flesh is flesh; and that which is born of the Spirit is spirit.*

In verse 3, Jesus stated that unless a person is born again, he cannot see the kingdom of God. The word again (G509) is better translated as "from above." Jesus was more specific in verse 5 when He said the person needs to be born of the Spirit, and that which is born of the Spirit is spirit. I Thessalonians 5:23 states:

> *And the very God of peace sanctify you wholly; and I pray God your whole spirit and soul and body be preserved blameless unto the coming of our Lord Jesus Christ.*

You are a whole that consists of three essential parts: spirit, soul, and body. Your eternal soul is what is being saved. This is the eternal you. Your intangible soul interacts with this world through your physical body, which includes your brain. Just as your brain is your soul's connection to the physical world, your spirit is your connection to the spiritual world. It is through your spirit that the Holy Spirit communicates into your brain. In order for the Holy Spirit to do this, your spirit must be born (activated). As for our analogy, in order for your car to get a signal to the GPS, your antenna needs to be activated.

I see everyone as having a spirit that is unborn. As Romans 10:9 states and we covered in *Modeling God*, when they confess with their mouth the Lord Jesus and believe in their heart that God raised Him from the dead, they birth their spirit through which they are supposed to get direction from the Holy Spirit via grace in order to grow in salvation.

Going back to our analogy, God's plan involves getting you to the party. The specific route God has for you is God's will, and it is done by following the specific directions God gives you. Hence, this book is focused on helping you understand God's will well enough to know if you and others are following their GPS or not.

Calvinism Comparison

God's will is a complicated topic, which is why about five hundred years ago, a simple answer to understanding God's will was widely accepted. Not only did this simple answer not align with the Bible, it is in opposition to the Bible, yet people have accepted it. While I'd rather not deal with this, the belief has so infiltrated Christianity that it is the biggest reason this topic has become more complicated.

One of the hardest things to do is unlearn. One example is that many "Christians" were made so uncomfortable by *Modeling God* because it required them to unlearn their traditional man-made beliefs that they attacked me rather than resolve their contradictions. The man-made belief is Calvinism, named after John Calvin. I'm going to give a brief overview of Calvinism in this chapter, and a more detailed explanation will be given later in Part I.

Calvinism: A Protestant theological system that has roots in the philosophical system known as determinism. Determinism is the doctrine that all events, including human action, are ultimately determined by causes external to the will of the individual. Some philosophers have taken determinism to imply that individual human beings have no free will and cannot be held morally responsible for their actions.

The belief in determinism has roots going back more than 300 years before Christ came to earth. This was not a belief presented in the Old Testament. Is this yet another example of a man-made tradition being put on the word of God that makes the word of God of none effect (Mark 7:13)?

In *Modeling God*, Dr. Joel Swokowski's Commentary made the following points when speaking about Calvinism as "the other explanation" from what the Bible stated:

The other explanation cannot support how a right and just God would create beings who were prevented from heaven and destined for eternal torture. Again, the discussion focuses completely on the explanation for this process because its existence is presupposed. The most destructive explanation is determinism: the future has already occurred and we are simply living out an existence that has been limited to only one possible day-to-day experience. Basically, God already knows what is going to happen and we not only don't have free will, we don't have freedom of choice. There are many passages that contradict this (Genesis 22:12: God saying "now I know..." when Abraham tries to sacrifice Isaac. Isaiah 7:15: Isaiah's prophecy about Jesus saying what He will do until He is able to choose the good and refuse the evil, etc.), however, the response is that we just can't understand it, which would make God unable to explain Himself to us.

The issue is that people can't separate out responsibility and credit. It looks as if everyone either believes God gets all the credit and is responsible for their salvation, or the individual gets all the credit and is responsible for their salvation. Why do we lump these two when it comes to God?

People tend to lump these two concepts, resulting in "experts" thinking there are only two possible explanations for salvation. If we get all the credit and have all the responsibility, then we are God. Clearly, this isn't a "Christian" explanation. If we get none of the credit and take none of the responsibility, then either everyone goes to heaven or God is unjust. This confusion between the concepts is where the two possible explanations that have caused division in Christianity occur.

Humanism/paganism believes we get **all** the credit and have **all** the responsibility.

Calvinism believes we get **none** of the credit and have **none** of the responsibility.

Fundamental Christianity, according to the Bible, believes we get **none** of the credit (it all goes to God), and we have **all** of the responsibility when it comes to our salvation.

The Biggest System

In the preface, we covered analytical and synthesis systems thinking. We saw the way to completely understand a system is to look at it from the perspective of the next bigger system. Dr. Russell Ackoff spent his life modeling business (or the corporation) by looking at it from the next bigger system: education. However, his ability to accurately model education depended on his ability to model the next bigger system.

In the true spirit of a synthesis systems thinker, he spent the last fourteen years of his life modeling the biggest system in order to be able to accurately model every system. He identified the biggest system as "God." Dr. Russell Ackoff believed the person who could model God would be able to model every system accurately, allowing them to repair every problem.

The issue is that God is not the biggest system. Remember, a system is a whole made up of essential parts. God is Holy, which means "of one substance." God does not have parts! God is not only the smallest system, God is the only entity in the universe that is not a system! This makes

God always existing and sufficient because nothing else existed outside of Him to support His creation or existence. In short, this proves God is the Creator. Everything would be created by God from outside of the created thing, just like a man creates a car. Notice, the car only works with a man as part of the system. Likewise, everything would only work with God being a part of every system He created.

If God is the smallest system, what is the biggest system? Albert Einstein said he wanted to know God's thoughts; the rest are just details. What Einstein was saying was that while God created everything for a purpose (His plan), He has a specific way He handles the parts of His plan. For example, how God allows bodies to move in space is guided by the system of physics. While, how God allows the human body to operate is guided by the system of biology. Einstein said all of these fields of study are lesser systems from the only system He cared about: God's will, that is, how God allows His plan to come about!

Notice, Calvin, like Ackoff, believed God is the biggest system. If God is the biggest system, He would be outside of everything, including His will, allowing Him to make anything happen because it would be a part of Him. This book believes God is the smallest possible system because He doesn't have any parts, while His will is the biggest system, a system that God Himself is a part of, just like a human is a part of operating and enjoying an automobile.

Let me be clear: God is sufficient. He doesn't have any parts and doesn't need anything in order to exist. However, God has a plan, which is a whole made up of essential parts, and God is one of the essential parts. My analogy is that God's plan is a party, and He very definitely wants to attend this eternal party. We will have to wait until the third and final edition to understand how that eternal plan works.

Part I of *Modeling God* showed how Jesus provided infinite value for salvation. Part I of this edition will show how God needs reward (spiritual value) to bring about His plan, which means God needs people to bring about His plan, AND God is sufficient.

Summary

- God's eternal party requires us to make progress toward an address, which is God's Nature (Right and Just).
- Our progress is measured by profitability.
- The best way to make progress is to allow God to lead you via grace.
- We are a mind/soul (ARE) driving a car (physical body), and grace is the GPS received through our spirit (antenna).
- Salvation is the door to the party.
- God's plan: The party itself is the ultimate in intimacy and communication accomplished by covenanting with others through our AREs.
- God's will is how He brings about the party through you.
- Understanding God's will allows you the ability to understand everything else in the universe.

Dr. Joel Swokowski's Commentary

Lenhart did a great job ensuring we understand the implications of what it means that God's will is the biggest system and how God Himself is the smallest. We'll see more and more of those implications as this book progresses. Allow me to join Lenhart in bringing some clarity and assurance.

God being the smallest system does not mean God is small. It is a result of God's holiness! It is an effect of God's purity. Something and/or some person that is made up of only one substance cannot have any impurities: no dust, no dirt, no darkness. The effect is they are separate or set apart.

None Effect

We know the Bible (the written words) is only truth when it's used with a right how/why. However, God's word is truth because God cannot help but use His words in truth (right how/why). So, why does the Bible have what seems to be an unending number of interpretations? Jesus gave the answer: "Making the word of God of none effect through your tradition…" (Mark 7:13a). In other words, man takes God's word and turns it into man's word.

God's word is truth. God's word is powerful. God's word brings profitability.

Man has a way of hindering God's truth, power, and profitability.

Tradition (habitual practices for the sake of habit) can stop God's word from bringing forth the benefits that God intended.

Here's a quick example. Let's say a church participates in communion (The Lord's Table) every Sunday. When the pastor proposes to limit this ordinance to once a month, they receive pushback from the elder board. Their reason: we have always done communion every week.

This gets to the heart of man-made tradition. We forget the God-given reason behind a behavior, and the focus becomes the behavior itself.

Communion at that church lost its power when the reason for doing that ordinance became: because we always have.

Tradition in and of itself is not bad. Man-made tradition that takes the place of the word of God is hindering the power of God, hindering His will, and hindering His eternal plan.

GPS

The GPS analogy is familiar and practical, at least, it is for me! My son Jack is a devoted volleyball player. He plays in two separate leagues over the twelve months of the year. For most of that time, he's playing for a travel league. We end up going out of town for the weekend at least once a month over eight months of the year. I use my GPS for every trip, to and from the city where the tournament is hosted, and all the driving in between.

As often as I use my GPS, as familiar as I am with it, the steps for using it remain the same. I put in my destination, and without fail, it asks me to "choose a starting point."

In order to get where I want to go, I need to know where I'm starting and where I'm located. As much as I want to go to Italy and bask in the sun of the Amalfi Coast, if I buy a ticket to fly there, I'm going to be required to tell them what airport I'm flying from. I need to know where I'm going and where I am currently located.

The first edition, *Modeling God*, gave us the tools we needed to determine where we're going (God), where we each uniquely reside (current beliefs/definitions), and how to intentionally and uniquely close the divide between the two (grace through our ARE).

This book helps us get a specific view of what going to God looks like and how to intentionally journey towards God's plan. The more we grow in our ability to hear from God, the more efficient and effective a path we'll be on towards God's plan. But don't fret, even if you do take a wrong turn, the GPS is quick to send you a re-routing message to get you back on the right path.

In fact, check this out from the end of "Chapter 5: Applications" of *Modeling God*:

How do you apply what we've learned in Part I in order to make progress toward God? In our party analogy, you would be constantly aware of how close you are getting to Righteous Street and Justice Avenue. Likewise, you must measure everything you do with respect to righteousness and justice. It is a process.

1. **Establish where you are today.**
   Recall from the analogy that the first thing I asked was where you were coming from. The sad truth is that most people don't know where they are coming from. They don't know what they believe. It's no wonder they can't make progress. Once you establish what you believe...

2. **Actively try to prove your personal beliefs wrong by identifying contradictions.**
   With respect to the analogy, this would be the same as figuring out if the street you are driving down is getting you closer to the party or further away. That leads to the final step.

3. **Change your beliefs in order to remove the contradiction.**
   That's it! That's the process of growing closer to God! However, you may decide you like the street you are driving on even though you

know it is taking you further from the party. Even though you are moving and know you are on the wrong street, progress still depends on your actions.

In order to make progress, you must choose to pursue growth instead of comfort.

Progress starts with determining where you are. A GPS won't work if you don't input your current location.

Do you know what you believe about God's will well enough that you can state it?

CHAPTER 2

God's Plan

SINCE THE TITLE of this edition is *Modeling God's Wills*, it would be crucial to not only clearly define God's will, but also clearly distinguish it from God's plan. The more you understand the difference between God's plan and God's will, the more quickly you will understand this complex topic.

God's plan is God's goal for eternity. God's plan is *what* God wants for eternity. Remember, *what* God wants, *why* God wants it, and *how* God goes about getting it must all be right and just. If we look at the end of the Bible, we see God's plan. Do you know God's plan?

Here's a simple question to ask: How does the Bible end? Can you give a simple explanation for the end of the Bible? You know, the way you summarize a movie in one sentence. How would you summarize the ending of the Bible in one sentence?

Most people would say, "Good people go to heaven, and bad people go to hell." Actually, both halves of that are wrong. If you don't believe me, read only the last four chapters of Revelation, which is the last book in the Bible. I know people think Revelation is hard to understand, but the last four chapters read like a movie script. Let's look at the key passages from the end of The Book of Revelation with commentary.

> *And the devil that deceived them was cast into the lake of fire and brimstone, where the beast and the false prophet are, and shall be tormented day and night for ever and ever. And I saw a great white throne, and him that sat on it, from whose face the earth and the heaven fled away; and there was found no place for them. And I saw the dead, small and great, stand before God; and the books were opened: and another book was opened, which is the book of life: and the dead were judged out of those things which were written in the books, according to their works. And the sea gave up the dead which were in it; and death and hell delivered up the dead which were in them: and they were judged every man according to their works. And death and hell were cast into the lake of fire. This is the second death. And whosoever was not found written in the book of life was cast into the lake of fire.* (Revelation 20:10-15)

The unholy trinity of the devil, the beast, and the false prophet are cast into the lake of fire and brimstone forever. Then everyone who has ever lived will face two judgments. There is a salvation judgment according to one book, the book of life, and determines the eternal destination of the individual. The other is a reward judgment, and rather than having one book like salvation, this one is done according to books. This judgment determines the amount of reward each person is due according to their works, resulting in spiritual value. Death, hell, and all those whose name is not found written in the book of life are cast into the lake of fire. Did you see that? Even hell is cast into the lake of fire!

We see the first adjustment to our one sentence conclusion for the Bible: "Good people go to heaven, and bad people go to *the lake of fire.*"

Modeling God spoke of these two judgments and showed that salvation judgment is done according to Right, the Righteousness of God through the individual by grace, while the reward judgment is done according to Just, the righteousness of the individual interacting with other individuals. Salvation is between the individual and God, while reward is between the individual and other people. Remember to keep these separate.

Let's look at the key verses in the next chapter of Revelation.

> *And I saw a new heaven and a new earth: for the first heaven and the first earth were passed away; and there was no more sea. And I John saw the holy city, new Jerusalem, coming down from God out of heaven, prepared as a bride adorned for her husband. And I heard a great voice out of heaven saying, Behold, the tabernacle of God is with men, and he will dwell with them, and they shall be his people, and God himself shall be with them, and be their God.* (Revelation 21:1-3)

While the last physical creation of God to date was woman (Eve), we see there will be three more creations in the future. God creates a new heaven and then a new earth because the current heaven and earth will pass away, or dissolve, that is, taken apart atom by atom.

The third creation will be the holy city, which is a new version of Jerusalem. The new Jerusalem is called the tabernacle of God because God will dwell with His people. Also, the new Jerusalem is compared to a bride adorned for her husband.

This implies a large part of God's plan is to live among people. Is that surprising? The Bible began with God personally interacting with Adam and Eve in the garden of Eden. Later, when God brought Israel out of Egypt, God said to Moses in Exodus 25:8: *And let them make me a sanctuary; that I may dwell among them.*

It has always been God's plan to live among a group of people who all feared/respected God, followed God, and sought His presence. God's plan was to only have these people present with Him for eternity. However, God the Father's story throughout the Bible was how certain people feared/respected, followed, and sought other gods, and this caused the rest of the people to not fear/respect, follow, and seek after God.

Notice, this passage from Revelation called the new Jerusalem the tabernacle of God, which is also God's dwelling place or house. What did Jesus say about His Father's house?

> *In my Father's house are many mansions: if it were not so, I would have told you. I go to prepare a place for you.* (John 14:2)

The new Jerusalem is approximately 1500 miles high, wide, and deep (Revelation 21:16). It is essentially half the area of the United States and equally high as it is wide and deep. If this structure were to be occupied by believers and each person's dwelling space was a half mile, by a half mile, by a half mile, there would be nine million units on the ground floor alone! Clearly, a living space of that size would dwarf any mansion we have on earth. Let's continue with this chapter.

> *He that overcometh shall inherit all things; and I will be his God, and he shall be my son. But the fearful, and unbelieving, and the abominable, and murderers, and whoremongers, and sorcerers, and idolaters, and all liars, shall have their part in the lake which burneth with fire and brimstone: which is the second death. And there came unto me one of the seven angels which had the seven vials full of the seven last plagues, and talked with*

> *me, saying, Come hither, I will shew thee the bride, the Lamb's wife. And he carried me away in the spirit to a great and high mountain, and shewed me that great city, the holy Jerusalem, descending out of heaven from God,* (Revelation 21:7-10)

The reason the new Jerusalem is compared to a bride adorned for her husband is that one of the seven angels called the new Jerusalem the Lamb's wife! Believers are the Bride and we will occupy the new Jerusalem in eternity, even though believers who die currently go to heaven.

Another way to see this is each person will make up a cell in the body of the Bride. The etymology for the word cell comes from Latin and means "storeroom" (Oxford Languages). So, in God's house, which is the Bride, there are many rooms or cells. Each believer will occupy a room that is easily the size of a mansion and then some! Just to be clear, here is the end of this chapter.

> *And the gates of it shall not be shut at all by day: for there shall be no night there. And they shall bring the glory and honour of the nations into it. And there shall in no wise enter into it any thing that defileth, neither whatsoever worketh abomination, or maketh a lie: but they which are written in the Lamb's book of life.* (Revelation 21:25-27)

We see the final adjustment to our one sentence conclusion for the Bible: "Good people *live for eternity in the new Jerusalem on the new earth*, and bad people go to the lake of fire." Was that your original answer for how the Bible ends?

Let's look at how the last chapter of Revelation began.

> *And he shewed me a pure river of water of life, clear as crystal, proceeding out of the throne of God and of the Lamb. In the midst of the street of it, and on either side of the river, was there the tree of life, which bare twelve manner of fruits, and yielded her fruit every month: and the leaves of the tree were for the healing of the nations. And there shall be no more curse: but the throne of God and of the Lamb shall be in it; and his servants shall serve him: And they shall see his face; and his name shall be in their foreheads. And there shall be no night there; and they need no candle, neither light of the sun; for the Lord God giveth them light: and they shall reign for ever and ever. (Revelation 22:1-5)*

The Bible ended with God, a groom (Jesus), and a bride (new Jerusalem) living near the tree of life with a river proceeding out from their residency. Sound familiar? Check out Genesis 2:7-10.

> *And the Lord God formed man of the dust of the ground, and breathed into his nostrils the breath of life; and man became a living soul. And the Lord God planted a garden eastward in Eden; and there he put the man whom he had formed. And out of the ground made the Lord God to grow every tree that is pleasant to the sight, and good for food; the tree of life also in the midst of the garden, and the tree of knowledge of good and evil. And a river went out of Eden to water the garden; and from thence it was parted, and became into four heads.*

Modeling God

As we showed in the previous chapter, we began *Modeling God* with the party analogy and we found the party represents God's plan for eternity. Here are quotes from *Modeling God* that referenced this.

From the beginning, I spoke about a party where you could experience any sensual pleasure you wanted to do for as long as you wanted. In order to try to approximate the level of joy you will feel in God's plan, I had to relate it to your current physical existence.

What God wants is to spend eternity with a growing sense of happiness and joy by interacting with beings who are completely enjoying themselves. The closest activity I could think of was a party where you get to do the activities you enjoy so much you never get tired of doing them; after all, this is an eternal party.

During the party, the activities you do for eternity will only result in more joy as you become better at, and focus on, them alone. God's plan requires people who desire to operate in their ARE for eternity. The only way this is possible is for each being to receive pleasure from operating in their ARE. In fact, some forms of these pleasures exist for us today.

Revelation 2:17 said:

> *He that hath an ear, let him hear what the Spirit saith unto the churches; To him that overcometh will I give to eat of the hidden manna, and will give him a white stone, and in the stone a new name written, which no man knoweth saving he that receiveth it.*

I believe the white stone will reveal to each person who they were created to be; their ARE. God's plan requires people to grow in who they were created to be in order to experience joy for eternity. Likewise, this will allow God to operate more in who He is.

This is God's plan. This is what it is all about. This is why we are here. This is the meaning of life! Your ultimate goal is to find and operate in your ARE. This is your calling.

When we think of God's plan, we ought to immediately think of God creating a community that will be completely focused on allowing who God created them to be to come out more, instead of focusing on a community that is limited by a law.

In eternity, the party will consist of beings in covenant with several other beings through their ARE. This will require the ultimate in intimacy and the ultimate in communication. Every being will know what they are valued for and how much they value every other unique individual. There won't be jealousy, because everyone will be perfect in his or her uniqueness.

Remember, according to systems thinking, the value is not in the individual, but in the exchange between individuals. Isaiah 9:6-7 says this:

> *For unto us a child is born, unto us a son is given: and the government shall be upon his shoulder: and his name shall be called Wonderful, Counsellor, The mighty God, The everlasting Father, The Prince of Peace. Of the increase of his government and peace there shall be no end, upon the throne of David, and upon his kingdom, to order it, and to establish it with judgment and with justice from henceforth even for ever. The zeal of the Lord of hosts will perform this.*

The only way for Jesus' government and peace to grow in eternity is that value needs to be continually generated, and this will be done by the interaction of beings in their uniqueness with the right and just Holy Spirit flowing through them via grace. This is God's plan!

If God's plan is *what* He wants, and the previous explanation is *why* He wants it, then God's will is *how* God accomplishes this plan.

Since this edition is about God's will, let's go much deeper with the next chapter.

Dr. Joel Swokowski's Commentary

What, Why, How

Lenhart stated that *"what God wants, why God wants it, and how God goes about getting it must all be right and just."* Believe me, he will go into detail to help you understand this. For now, here's a basic understanding of the three:

WHAT: This is the effect we see of "Who is God?"
God's Nature is Righteous and Just.

WHY: In other words, "Why is He able to be Righteous & Just?"
God is able to be Righteous and Just because it is in His Nature: always & completely.

People tend not to argue about God being Righteous and Just, until you get to this part. When talking about God's Nature (His *why*), this is where the contradictions are exposed!!

HOW: In other words, *"HOW is God able to be Righteous & Just?"* ...

...or in other-other words, "How does God ensure it's *impossible* for Him to be wrong/unjust?"

What did God do on the seventh day of creation? He rested. This means He ceased from the occupation of being a first cause. God is in a state of response. He moves for, against, or not at all *in response* to Justice.

Paradise

In the last few years, the author and some other friends of ours who are leaders in church met a group of people in Scotland who were on a journey of discovering truth. The journey they have been on has inspired awe in us, and we've been blessed to be part of their lives. Recently, a small group of us were able to travel to Scotland to help them start a church. Part of this trip included leading a baptism in the River Tay, in Scotland…in February!! It was incredible, I get chills and a warm heart thinking of it still.

I was honored to participate in the baptism as I watched some new friends express their faith in Jesus through water baptism, in the River Tay, in Scotland…in February!! I look back at that event, and I can tell you, I was completely immersed in the experience. As cold as the water was, I was entirely focused on watching people communicate their love for Christ through this special ordinance. As emotional as I was, and moved to an intense excitement, I was wholly captured by the moment, watching the leaders of Relight Church do the work of the Lord. There was nothing pulling my thoughts to the past or pushing me to imagine anything in the future. I was absolutely in the moment; there was nowhere else I would have rather been…in the River Tay, in Scotland…in February!!

We're heading towards an eternal party. This is contingent on God creating a situation that always grows and is eternally profitable—eternal paradise!

What is the definition of paradise?

When you are doing something that is *not* the thing you most want to do, you tend to think about something in the future. You are demonstrating faith. You are demonstrating hope. You are believing and thinking about something that hasn't happened yet.

When you are doing the thing you *most* want to do, you *don't* think about the future. You *don't* demonstrate faith. You *don't* demonstrate hope.

When we are doing the thing we enjoy *most*, our thoughts are *all* in the present. We aren't looking forward or backward.

That is paradise. Time continues; however, we are not thinking about the future because we are doing the thing we *most* want to do. Love (charity) and time would continue; they wouldn't fail. However, faith and hope would end.

> *And now abideth faith, hope, charity, these three; but the greatest of these is charity.* (I Corinthians 13:13)

CHAPTER 3

God's Contrastive Process

WHILE GOD'S PLAN is the only plan that would work for eternity, it would be wrong and unjust for God to begin with that plan. If He did, someone could say, "I think God could have accomplished this by giving us a set of rules and threatening us with an eternal punishment if we didn't do them." Even though we know that wouldn't work, it would prove God wasn't right and just because He would be initiating His will on everyone, which violates everyone's free will. God's will first involves God justly proving His plan is right by using a contrastive process.

God's plan is right and just AND the process God uses to bring about His plan is right and just. I call this Right-Right to remind people God isn't just right in *what* He does, He is also right in *how* and *why* He does it. God's right and just method is to show that every other possible configuration is wrong and that every other configuration ends up unprofitable and unable to continue to grow for eternity. These configurations are called dispensations.

The dispensations are:

1. Eden - completed
2. Pre-Flood - completed
3. Post-Flood - completed
4. Abraham - completed
5. The Law - paused
6. Pentecost - current
7. Millennium - yet to happen
8. God's Plan - yet to happen

Each dispensation has different conditions. People tend to see contradictory behavior from God when they compare between dispensations. The differences are the ethics. They are correct for the dispensation only and help us understand God's objectives for each dispensation. The similarities across dispensations are God's morals.

In *Modeling God*, we saw that ethics are principles that result in actions that a specific society living at a specific time deems as allowable, even though they are not profitable. Morals are principles that result in profitable actions regardless of the society and the era.

The Bible documented all eight dispensations, with God's plan (#8) being the only way for people to exist profitably forever. God's eternal plan is the only one that is able to continually grow in profitability for eternity. The order of the dispensations is important because God couldn't undo certain conditions. God also accounts for the flaws in a dispensation by addressing it in the ensuing dispensation. We will also see the need for thinking increases with each dispensation.

Eden

Eden is the first dispensation. The idea behind this dispensation was: What if people weren't distracted by worrying about good and evil, but instead, they had a direct connection with God, so they could just do everything God said? Since they didn't know evil existed, they wouldn't have guilt.

At first blush, this sounds like a perfect way to live for eternity. However, God couldn't oppose their will, so He had to give them the ability to reject His plan. God gave them access to the tree of the knowledge of good and evil. This lets them express their will by not eating from it.

In *Modeling God*, we saw that faith is based on understanding and experience. Furthermore, understanding (the *why*) is based on knowledge (the *what*). Adam and Eve had little to no experience with God, so all they could have relied on was understanding (which is based on knowledge). The whole point of this dispensation centered on them not having knowledge and instead taking complete direction from God! They didn't know the serpent was evil. They didn't know what "evil" meant. This dispensation would never work.

Notice that even though we now understand *why* this dispensation would never work, God had to create the scenario to prove it was flawed. As for the order of the dispensations, this had to be the first because people couldn't forget about the knowledge of good and evil since we are descendants of Adam (Romans 5:12).

Pre-Flood

The next dispensation was between Eden and the flood. This began in Genesis chapter 4 and ended in chapter 7. The first dispensation failed because people lacked faith. They were unable to believe in something they couldn't see or hadn't happened yet. Their actions were only good when God was visible.

In the second dispensation, God allowed people to live hundreds of years so they were able to build their understanding (thinking) and experience of God. However, these people had no sense of urgency! The result was they only got more focused on the flesh. The Bible said Noah was the only righteous person on Earth (Genesis 6:8-9).

This was the right time for this dispensation. If God was ever going to let people live hundreds of years in only the naturally habitable areas, it had to be done while the world population was low.

Post-Flood

The third dispensation occurred after the flood. This dispensation covered Genesis chapters 8 through 11. The previous dispensation failed because people had no sense of urgency, so the lifespan was shortened in this dispensation from about 800 years to (eventually) about 80 years.

God also had to justly shorten the lifespan of all people equally. No one had an advantage over others. In order to do this, God removed everyone from the previous generation at the same time to justly transition into the next dispensation.

The Bible said the people of this third dispensation focused on thinking (Genesis 11:6). In fact, they were able to do anything physically because they spoke the same language and were of one mind! However, their focus was on the flesh instead of God.

Remember, God wants to dwell with people. In order to facilitate this, God mixed the languages at the Tower of Babel to help them stop focusing on the physical. This also created nations.

So far, we've seen three dispensations, and they all resulted in people focusing on the tangible instead of God. In the first, it was because they didn't know any better. In the second, it was because they had no sense of urgency. Here, in the third, it was because they were physically able to do whatever they imagined.

Abraham

The fourth dispensation revolved around God picking a specific person and using him and his descendants as a teaching example for the rest of the world. This could only have happened after nations were created. This dispensation began with God forming a covenant with Abram in Genesis chapter 15.

The rest of Genesis was really an in depth look at how God tried to make other nations good by demonstrating the principles of Right and Just through His interactions with this one nation. For example, Abraham still had to prove his focus was faith in God and not on the physical by being willing to sacrifice his son, Isaac (Genesis 22:12). This allowed God to continually bless Abraham.

The story of this family goes from Isaac to his son Jacob, and then from Jacob to his sons. Along the way, we encounter almost every kind of evil imaginable that can occur between people. Anyone who reads Genesis will realize Abraham and his relatives were a very dysfunctional family that were still blessed by God because of their faith.

Clearly, this dispensation didn't make God's people good. As for the rest of the nations, it only made them jealous. These nations focused on wiping each other out throughout the rest of the Bible.

Clearly, the fourth dispensation didn't result in a community of people being good. In fact, it appeared that some of the people in this fourth dispensation didn't recognize their actions as sinful. Did these people objectively know what sin was?

Genesis covered the first four dispensations. The rest of the Old Testament and the Gospels covered the fifth dispensation.

The Law

The overwhelming majority of the Bible dealt with the fifth dispensation. After Genesis, the next four books of the Bible documented the liberation of Israel and the creation of the law. Paul said the purpose of the law was so that we would know sin. (Romans 7:7) The problem with the fourth dispensation was they didn't objectively know what was sin.

God set up this fifth dispensation the way most people would set up a society. He gave people objective rules and punished them severely when they broke the rules. That seems intuitively obvious to us. However, why do we believe this approach makes people good? Does it make people good? If not, why do people naturally resort to this approach?

We all know the law was written in terms of justice. One of the most famous quotes from the law was "eye for eye" (Exodus 21:24). The law tried to enforce justice on earth by stating the penalty for each offense.

However, the Old Testament was really proof that the law doesn't work. It doesn't make people good. It doesn't cause them to focus on God. It gave a reason *why* things went bad for them and motivated people to turn back to God. However, as long as things were going well, people tended to take their focus off the spiritual and put it on the physical.

Remember, in the big picture, God was proving to us none of these dispensations work. The law didn't work. It doesn't make a group of people good. It doesn't cause a community of people to focus on God more. All the law really did was prevent chaos. If you expect anything more than that, you will be sorely disappointed.

Clearly, while the law itself was good, it was ethical in that it only applied to this dispensation. We are not under the law currently. God didn't hold people to the law in the dispensations previous to the law. However, even today, people try to claim the law is moral. That is, they try to claim that God uses it as the measure for every culture and at every time in history. The problem is there are several examples where God honored people who broke the law. Some quick pre-law examples include Moses and Abraham.

Moses killed an Egyptian and then felt guilty enough to run away (Exodus 2:12-15). Yet God treated Moses as a friend (Exodus 33:11). Think about it, the person whom God chose to present the Ten Commandments was a murderer. How can anyone say the Ten Commandments are moral? Moses killed the Egyptian before the law was given. Are the Ten Commandments only true after God stated them?

Abram married his father's daughter, which was contrary to the law (Genesis 20:12). Yet, God formed a covenant with him. Again, how can anyone say the law is true across all dispensations, (i.e. moral) if God didn't hold it against Abraham?

Pentecost

The Book of Acts introduces our current dispensation. Before the cross, Jesus was fulfilling the law, so He was living in the dispensation of the law. The next (our current) dispensation actually began at Pentecost through the Holy Spirit.

Before we discuss our current dispensation, there's one tricky point to understand. Jesus fulfilled the law, so things He said and did before His death could be for that dispensation and were, therefore, ethical-based. For example, Jesus paid the temple tax in order not to offend those who collected it (Matthew 17:24-27). Jesus was also moral, so some of the things He said could apply for any time. Finally, there were moments when Jesus spoke of the future. Depending on how He introduced it, those comments either were intended for only the future dispensation or were moral and applied to all.

In order to accurately identify the application of Jesus' words and actions, pay close attention to whom Jesus was talking and what side of the cross you are observing. Remember, before Jesus was crucified, He was fulfilling the law. In fact, one of the purposes of the law was for us to know for sure that Jesus is the Christ (Romans 10:4)! If Jesus didn't fulfill the law or taught that the law resulted in sin, then it would prove to us that He wasn't the Son of God.

For instance, Jesus would not tell us that following the law led to sin. This is different from not following the law. Think of the woman caught in adultery in John 8. When she was brought to Jesus, He was reminded that the law said to stone her. Then Jesus was asked what He thought ought to be done. If Jesus had said, "Stoning her is a sin," then Jesus would be saying that doing the law results in sin, and this would prove to us He wasn't the Son of God.

Jesus said that he without sin ought to cast the first stone. This dispersed the crowd. Notice, the law was not followed and this wasn't a sin. However, Jesus would never say that doing the law resulted in sin.

The overwhelming majority of the Bible documented the dispensation of the law. People tend to see the entire Old Testament as under the fifth dispensation. They tend to associate the entire Old Testament, and only the Old Testament, with the law. We've seen that the dispensation of the law covered all of the Old Testament except Genesis (and Job). Notice, from this dispensational perspective, even the four Gospels are technically Old Testament books because they occur during the fifth dispensation.

The dispensation of the law began with God wanting to speak with everyone. Remember, God wants to dwell with everyone. Yet the people told God to talk to Moses (Exodus 20:19). God's response was to give the law. The law told people what not to do. All this did was cause people to want to do what they ought not do.

Our current dispensation is grace (charis) through the Holy Spirit. The Greek word charis has been translated into several English words: favor, gift, thank, etc. by experts depending on usage. When it is translated into grace, its definition, according to Strong's Concordance (G5485), is "the divine influence upon the heart, and its reflection in the life." *Modeling God* covered this in depth.

Jesus is speaking to everyone through the Holy Spirit. Where the law told people what not to do, Jesus is telling people what He wants to do through them. He can do these things through them if they choose to allow Him to flow through them. If their spirit is born, they are able to hear His direction. If their spirit is not born, they are unable to hear His direction; however, Jesus continues to speak to them because He is right and just. It is not His fault the person didn't choose to birth (activate) their spirit.

Our dispensation is failing because people don't understand grace (divine influence upon the heart, and its reflection in the life). People are actually stuck in the fifth dispensation and making the word and will of God of none effect by our tradition (Mark 7:13).

Our current dispensation (Pentecost; grace) ends after the Rapture of the church. However, instead of the introduction of the seventh dispensation, the fifth dispensation returns! The antichrist is introduced, and the world goes through seven years of tribulation under the law.

This is spelled out in the books of Revelation and Daniel. This is the seventieth week Daniel spoke about in Daniel 9. God re-establishes the covenant with Israel, and from this point on, it can be kept forever.

Millennium

At the end of the Tribulation, the Antichrist is removed and Satan is locked up for 1000 years. This period marks the seventh dispensation. Notice, this seventh dispensation is a sabbath (time of rest) from the influences of the devil.

The Bible said God will be focused on nations again. Notice our current dispensation is one of the few where God doesn't honor nations. God is currently focused solely on individuals through the Holy Spirit. The Bible says God is still focused on individuals in the seventh dispensation. However, the Bible also speaks about the nations coming up to worship Jesus. If they don't come up to worship, their nation will get no rain (Zechariah 14:17).

There will be peer pressure within nations, but no conflicts between nations. The Bible said it will be a time of peace. The most famous quotes about this dispensation are "the wolf also shall dwell with the lamb" (Isaiah 11:6), and "they shall beat their swords into plowshares" (Micah 4:3, Isaiah 2:4). Also, Isaiah said people will live a long time, to the point that those who die at one hundred years old will be considered children (Isaiah 65:20). Notice, it may be the only way the earth could support a population that lives hundreds of years would be for the population to be relatively low. The Tribulation results in a severe reduction in the population of the earth.

It is unclear whether the group that helps out with the government of these people during this dispensation are only Christians who have been martyred or all people who have been previously saved. This is God's way of addressing the problem with our current dispensation: we don't follow the Holy Spirit! Instead of intangible grace from within through the Holy Spirit, the people of the seventh dispensation will be influenced by tangible people (already saved) coming back to rule and reign with God. According to our analogy, instead of being led by an intangible GPS signal, everyone will be led by a tangible co-driver.

At the end of the thousand years, Satan will be released and he will make one last attempt at God. The battle will be over quickly and the dispensation ends, but not before Satan has deceived even more people. Apparently, people will still choose the flesh. This dispensation will be a failure as well. Why would a time of peace, long life, and rest from satanic influence fail?

Because the devil is locked up, people live long without poverty, sickness, etc. This sounds ideal. In fact, people living in our current dispensation think our problem is the devil. They say they wish there were no devil so there would be no problems. However, we now know God is showing us that the dispensations don't work.

When people talk about how they wish the devil were gone, I ask them, "How would you help people focus on God and grow?" Currently, people rely on the negative effects in order to get people to focus on God. It seems the way we evangelize today is to find people who are sick, poor, depressed, or afraid of hell and tell them about Jesus. Current evangelism attempts to get them across a salvation finish line.

It seems Christians currently need the devil in order to convert people because they don't understand the *why* and/or the benefits of grace. They don't define the benefits. They define the benefits by what they are not, as if a lack of problems is a benefit. Today, some people define good behavior by lack of sin.

People will have guilt in the Millennium, but there will be no condemnation because the devil is absent. Without condemnation, it is very difficult to feel a sense of urgency for the individual to confess and repent. Also, since these people will live long lives without trials and tribulations, they could think they don't need anything. They may not think they need the grace provided by the martyrs.

In fact, I sometimes call this dispensation the "Solomon Dispensation." These people are going to have a relatively easy life and have all the knowledge of every preceding dispensation. They will be the wisest dispensation while lacking trials and tribulations, just like Solomon.

What if people in the Millennium realize they have hundreds of years to live and plenty of time to make a profitable decision? They may decide to wait hundreds of years before they express their will towards God.

Just like Solomon, they may understand the situation, but their will is proven by their actions. They need to do the process.

However, when people spend hundreds of years ignoring grace, they may sear their conscience over and it would have little to no effect on them. It looks as if we may actually be relying on the devil currently for evangelism because he accentuates guilt!

One final point: Notice that while none of these dispensations result in an entire group of people choosing to pursue God, it is possible for individuals to choose to fear/respect, follow, and pursue God. Not only does this prove who actually wants to spend eternity with God, but the process for their salvation is the same in each dispensation: by grace through faith. The difference in each dispensation is the method for the divine influence upon the heart. For example, in Eden it was directly from God, while during the law it was from His word, and today it is the indwelling Holy Spirit.

As we've seen, the only way for a community of people to live profitably for eternity is God's eternal plan because all the other options don't work. In order to appreciate this edition's explanation for God's will, we need to look in depth at the current explanation being used today.

Dr. Joel Swokowski's Commentary

Jesus' Contrastive Process

We've learned about contrastive thinking in *Modeling God* and how it is the only way to prove something is truth. For example, Jesus prayed contrastively when He didn't want to go through the specific process that was laid out to Him.

> *And he went a little further, and fell on his face, and prayed, saying, O my Father, if it be possible, let this cup pass from me: nevertheless not as I will, but as thou wilt.* (Matthew 26:39)

Jesus asked if it was possible to be crucified under different circumstances.

> *And he cometh unto the disciples, and findeth them asleep, and saith unto Peter, What, could ye not watch with me one hour? Watch and pray, that ye enter not into temptation: the spirit indeed is willing, but the flesh is weak. He went away again the second time, and prayed, saying, O my Father, if this cup may not pass away from me, except I drink it, thy will be done.* (Matthew 26:40-42)

Jesus prayed contrastively. He prayed, "if this *may not* pass away." Either way, Jesus was willing to do God's will, God's way, even though He did not want to.

Rapture of the Church

Lenhart mentioned that our dispensation of grace ended with the rapture of the church. In I John 5:18, it is taught that anyone born of God cannot be touched by the wicked one. The power of the Holy Spirit within believers trumps any power that the devil or his minions carries. This means the antichrist would be no match for anyone with the Holy Spirit, so the Holy Spirit must be gone during the seven-year period known as the Tribulation. This would coincide with the dispensation of the law coming back with the seventieth week of Daniel.

There is a long-standing debate about when *the* rapture happens: pre-, mid-, or post-Tribulation. The first problem is the belief that there is only one rapture. If you wanted to be technical, we could go all the way back to Genesis to see the first recorded rapture: Enoch (Genesis 5:21-23, Hebrews 11:5).

For the sake of our point here, regarding the Tribulation, the three raptures argued about are all valid. Here is a brief description of each of the three Tribulation raptures:

1. Pre-Tribulation: Removes those born of God from earth before the Tribulation in order for the seventieth week of Daniel to begin (Revelation 4:1).
2. Mid-Tribulation: Removes all the people who were converted, without having the Holy Spirit, during the first half of the Tribulation, so they avoid the second half of the Tribulation (Revelation 7:9-10).
3. Post-Tribulation: So that all the souls that are on Jesus' side are gathered together when the Two Witnesses are raptured for Armageddon (Revelation 11:12).

Achieve Gain vs Fear of Loss

In *Modeling God*, Lenhart taught about the two ways to motivate a person to grow: achieve gain or fear of loss. In the commentary of "Chapter 7 Faith Examples," I wrote:

The born-again experience is a very emotional and transformational moment for everyone who goes through it, and rightly so. If you're a Christian, you know for yourself and you've seen new Christians seem to have a youthful zeal that is attractive, often eliciting envy in the more seasoned believer. There comes a problem though, not long after conversion if the person who's become born-again doesn't add understanding to their foundation, to their newly adopted Christian worldview. If something challenges their faith and that faith is mainly experiential, it's likely that this person could fall away. If that person is given a strong foundation of understanding, it can help the person deal with the challenges that come their way, specifically the "bad" experiences, and the person can remain strong in their faith regardless of the context.

In this chapter, Lenhart showed how much of the church today uses the devil as an evangelical tool. This tool has motivated people to get saved in order to not go to hell. This is a fear of loss approach to salvation (growth). Now, there's nothing wrong with using this method to evangelize nor is there anything wrong if you were saved for this reason.

There is, though, a long-term issue with this approach. What happens when your fear is alleviated? If you are assured of your salvation and no longer fear hell, what benefit are you going towards that will be your motivator for growth? Having the correct doctrine about the salvation process gives you and any person you lead to Christ the ability to choose the achieve gain approach, choosing to grow because it's profitable, even if you began with the fear of loss perspective.

CHAPTER 4

Calvinism

HOW DOES GOD bring about His will?

In *Modeling God*, we saw how philosophers described God's Nature with three "omnis": omnipresent, omniscient, and omnipotent. To the philosophers, omnipresent means God is everywhere, omniscient means God knows everything, and omnipotent means God is all-powerful and can do anything. These three omnis are summarized with the doctrine of sovereignty.

Again, we saw in *Modeling God* that sovereignty means doesn't answer to anyone. However, the philosophers wrongly concluded sovereignty means that God can do anything and we can't understand it, while we showed how the Bible contradicted this explanation for God's Nature. For example, the Bible stated in two separate places that it is impossible for God to lie (Hebrews 6:18, Numbers 23:19).

Unfortunately, Christianity embraced this contradictory philosophical belief, which is the basis of the most popular explanation for God's will. This man-made belief is known as Calvinism and consists of five doctrines known as "The Five Points of Calvinism." These doctrines are named for the distinct theological stances taken by the Protestant reformer John Calvin (1509-1564).

In Chapter 1 of this book, we mentioned that this Protestant theological system has roots in the philosophical system known as determinism: the doctrine that all events, including human action, are ultimately determined by causes external to the human will. If you've heard of The Five Points of Calvinism, you have likely encountered the acrostic T.U.L.I.P. as a memory aid for these doctrinal positions.

The Five Points of Calvinism are:

> **T** – Total depravity: Man is completely sinful.
>
> **U** – Unconditional election: God chose ahead of time who He would "elect" as saints/believers.
>
> **L** – Limited atonement: Jesus' death provided salvation for the elect only.
>
> **I** – Irresistible grace: The elect cannot resist God's will to save them.
>
> **P** – Perseverance of the saints: If you have been justified before God, you cannot lose your salvation. Once a person is truly saved, this salvation is eternally secure. Also known as "Once saved, always saved."

Ephesians 1:9-12 explained that the mystery of His will has been made known to believers and is a great summary for everything we will discuss:

> *Having made known unto us the mystery of his will, according to his good pleasure which he hath purposed in himself: That in*

the dispensation of the fulness of times he might gather together in one all things in Christ, both which are in heaven, and which are on earth; even in him: In whom also we have obtained an inheritance, being predestinated according to the purpose of him who worketh all things after the counsel of his own will: That we should be to the praise of his glory, who first trusted in Christ.

The word that people focus on is predestinated or, in other versions of the Bible, foreordained. In order to go forward in a focused fashion into more complex areas, we need to address this one word that tends to distract people.

Predestination

When each person is conceived, God can see the gifts and talents of the person relative to the time in history when they are born. So before we are formed in the belly, I believe God knows us and predestines works for us to achieve. We saw these as opportunities to gain reward (spiritual value), if we follow the GPS during our journey to the party. We see this in the next chapter of Ephesians with the usage of "before ordained":

For by grace are ye saved through faith; and that not of yourselves: it is the gift of God: Not of works, lest any man should boast. For we are his workmanship, created in Christ Jesus unto good works, which God hath before ordained that we should walk in them. (Ephesians 2:8-10)

As we saw in *Modeling God*, the Biblical definition of predestinate is "to limit in advance." For example, men are predestined not to bear children. They are limited in advance. We are all limited in advance to where we can ultimately end up for eternity: new Jerusalem or the lake of fire. We are all predestinated.

Some people who believe in the Calvinistic view of predestination abuse this definition to say that every second of our life has been limited in advance to only one possible choice and/or outcome. For instance, God knows exactly where you will be and what you will say at noon on your thirty-seventh birthday. This is taking the concept farther than it was intended, and the resulting contradictions that pile up due to all the free will, responsibility, and judgment verses in the Bible testify against people who hold to this man-made, traditional definition.

Believing God knows everything that will ever happen and is all-powerful results in determinism because it denies every person's free will. Determinism believes God sees every moment as occurring at the same time, which means the future has already happened and we don't have free will. C.S. Lewis showed in his book *The Discarded Image*, that this doctrine was mentioned in Boethius' (480-524 AD) fifth book *De Consolatione Philosophiae*, and not from John Calvin. However, notice this was not a belief held by anyone in the Bible. How do Calvinists try to support their position with the Bible?

Romans 8:26-30 is a favorite passage for people who hold to the man-made predestination and/or foreordained definitions:

> *Likewise the Spirit also helpeth our infirmities: for we know not what we should pray for as we ought: but the Spirit itself maketh intercession for us with groanings which cannot be uttered. And he that searcheth the hearts knoweth what is the mind of the Spirit, because he maketh intercession for the saints according to the will of God. And we know that all things work together for good to them that love God, to them who are the*

called according to his purpose. For whom he did foreknow, he also did predestinate to be conformed to the image of his Son, that he might be the firstborn among many brethren. Moreover whom he did predestinate, them he also called: and whom he called, them he also justified: and whom he justified, them he also glorified.

This passage began with how the Holy Spirit makes intercession for us according to the will of God. Since God knows the Mind of the Spirit, God therefore knows what is being planted in our hearts if we let the Spirit flow through us (grace). It would be unjust of God to condemn our imperfections if there was no way to overcome them. Grace allows us to overcome our imperfections by choosing to let God work through us. This ability isn't something we can take credit for because it actually requires us to allow God to do it through us (grace).

The next part of the passage talked about how all things work together for good to them who love God and are the called according to His purpose. We saw in *Modeling God* that good means to create or be profitable, not necessarily perfect. The perfect result is when we flow in grace. If you don't follow grace, the passage states God is still able to make the situation profitable instead of a loss, if you love God and follow His purpose.

The end of the passage has been massively abused because it is interpreted from effect to cause. The abusers say God foreknew everyone, so He glorified everyone, which means everyone is saved. However, the truth is whether God glorifies the individual or not, is up to the individual!

This passage is correct in saying that everyone who is glorified (saved) was foreknown, predestinated, called, and justified; however, not everyone who is foreknown, predestinated, called, and justified is glorified. Reread the passage. We have a choice whether we are glorified (saved) or not. This is a perfect example of how people abuse causality to prove their point, whether they intend to or not. It is subtle.

For example, *Modeling God* looked at how God hardened Pharaoh's heart as an effect of Pharaoh's decisions and was not against Pharaoh's will. Is there any place in the Bible where this interaction with Pharaoh was seen as Pharaoh unable to express his will? Actually, the opposite occurred!

In I Samuel 6, the ark of the covenant ended up in the land of the Philistines. Here was what the Philistines said about it:

> *Wherefore then do ye harden your hearts, as the Egyptians and Pharaoh hardened their hearts? when he had wrought wonderfully among them, did they not let the people go, and they departed? Now therefore make a new cart, and take two milch kine, on which there hath come no yoke, and tie the kine to the cart, and bring their calves home from them: And take the ark of the Lord, and lay it upon the cart; and put the jewels of gold, which ye return him for a trespass offering, in a coffer by the side thereof; and send it away, that it may go.* (I Samuel 6:6-8)

The Philistines recognized Pharaoh had a free will and made the wrong choice, so they expressed their free will to make the right choice. Do Calvinists know less about God's doctrine than the Philistines? The passage we looked at that explained salvation was Ephesians 2:8-9.

> *For by grace are ye saved through faith; and that not of yourselves: it is the gift of God: Not of works, lest any man should boast.*

Do you see that these are salvation verses?

Let's look more closely at the passage that immediately followed those verses.

> *For we are his workmanship, created in Christ Jesus unto good works, which God hath before ordained that we should walk in them.* (Ephesians 2:10)

Do you see that this is a reward verse?

This verse mentioning good works was not focused on salvation, otherwise it was not salvation according to the verses that immediately preceded this passage. It was focused on opportunities for rewards that God had placed before us that we choose to walk in through God's direction (grace).

God has a perfect route (His will) for our life based on all the information that is available at that moment. Remember (from *Modeling God*), the non-contradictory definition of perfect is "maximum profitability." This perfect route from God will result in a finite reward, and again has nothing to do with your salvation. This finite reward is different from what others can achieve, both more and less.

The Bible stated the following about Jeremiah in Jeremiah 1:5:

> *Before I formed thee in the belly I knew thee; and before thou camest forth out of the womb I sanctified thee, and I ordained thee a prophet unto the nations.*

This doesn't necessarily mean that God knew us before we were formed, but it wouldn't be contradictory to believe that if God knew Jeremiah, God could also know us.

However, it would be contradictory for God to know anyone other than Jesus before they were *conceived* because Jesus was already begotten. People use this verse from Jeremiah to say it proves not only that God knew all of us but that He knew us before we were conceived. This verse didn't state either of those conclusions. Also, God would know our ARE (the causes), not necessarily our personality, which is an effect of our circumstances, including the actions of other people. Be careful!

"Formed" in the above passage came from H3335 "yatsar" meaning "to mold into a form; as a potter" and "to determine (i.e. to form a resolution)."

So, we have the consistent imagery of God having an objective (knew thee) before He molded Jeremiah (and probably all of us) into the form that is perfect for who we ARE. It would be wrong of the potter to determine the exact dimensions of the pottery before the lump of clay was provided. If there is too little clay, then the potter doesn't make exactly what he knew. If there is too much clay, then the potter makes something less than what he could have made. God is right and just. He would determine the purpose once the clay was provided (conceived), once God gave us our ARE.

People who are able to achieve more than others are given more from God than He gives to others: God loves them more.

People who are given less are loved less or, according to the King James Bible, they are hated.

God loves everyone; God gives everyone the opportunity to gain reward. God's plan for everyone involves dwelling with Him in eternity in the new Jerusalem. However, the minute we use comparative words (more, less) we need to adjust our mindset. Being hated by God doesn't mean the individual is not loved at all. It just means they are loved less. Look at Genesis 29:30-31.

> *And he went in also unto Rachel, and he loved also Rachel more than Leah, and served with him yet seven other years. And when the Lord saw that Leah was hated, he opened her womb: but Rachel was barren.*

Notice, Jacob loved Rachel more than Leah, and God said that Leah was hated. If you don't like this definition, talk to God. Notice also, man-made predestination traditionalists like to quote "Jacob I loved and Esau I hated" like God didn't love Esau at all when a passage concerning Jacob showed this predestination interpretation is a contradictory and wrong interpretation of the word hate. However, there is more to this verse that makes the Calvinist wrong.

> *As it is written, Jacob have I loved, but Esau have I hated.* (Romans 9:13)

Paul specifically said he was quoting a passage from the Old Testament (it is written). The ultimate meaning of the verse ought to come from the context in the Old Testament! Where was this passage in the Old Testament? Here is the very beginning of The Book of Malachi.

> *The burden of the word of the Lord to Israel by Malachi. I have loved you, saith the Lord. Yet ye say, Wherein hast thou loved us? Was not Esau Jacob's brother? saith the Lord: yet I loved Jacob, And I hated Esau, and laid his mountains and his heritage waste for the dragons of the wilderness. Whereas Edom saith, We are impoverished, but we will return and build the desolate places; thus saith the Lord of hosts, They shall build, but I will throw down; and they shall call them, The border of wickedness,*

and, The people against whom the Lord hath indignation for ever. And your eyes shall see, and ye shall say, The Lord will be magnified from the border of Israel. A son honoureth his father, and a servant his master: if then I be a father, where is mine honour? and if I be a master, where is my fear? saith the Lord of hosts unto you, O priests, that despise my name. And ye say, Wherein have we despised thy name? Ye offer polluted bread upon mine altar; and ye say, Wherein have we polluted thee? In that ye say, The table of the Lord is contemptible. And if ye offer the blind for sacrifice, is it not evil? and if ye offer the lame and sick, is it not evil? offer it now unto thy governor; will he be pleased with thee, or accept thy person? saith the Lord of hosts. (Malachi 1:1-8)

The Malachi passage was between God and Israel; between a father/master and a child/servant. God was pointing out that He was not the cause of Israel's problems! In fact, God was saying that Israel was interpreting this verse the same way as the Calvinists and it is wrong!

God was showing Israel that they should rejoice they weren't Esau because Esau got judged and wasn't humble enough to respond correctly. However, everything God said about Esau (Edom) was an effect of Esau's choices (cause). Edom was going to build apart from God's help, and God was going to judge their pride and wickedness.

Then God brought the discussion back to Israel to say they weren't acting like His children; the children who were His effects of the covenant! Israel was in danger of becoming like Esau if Israel made the same choices as Esau even though God loved Israel (Jacob). If it all had been determined ahead of time, why would God be concerned for Israel's future?

The Malachi section concluded with the understanding that God wasn't going to bless Israel just because of who they are. They have to pray, admit they were wrong, and choose to do the right thing!

Does it concern you that Calvinists base their theology on the same wrong understanding that God warned Israel about?

In order to understand what God's word said about how God brings about His plan, we need to dig deeper into the Bible. The next chapters will begin this process.

Dr. Joel Swokowski's Commentary

The Fourth Omni

As we've seen, the three "omnis" (omnipresent, omniscient, and omnipotent) are the man-made basis of what most people believe is God's Nature. We've also seen how these contradict God being always and completely Righteous and always and completely Just. The three "omnis" are not the cause of who God is; they are not what makes God, God. I wouldn't be God if I were omnipresent, omniscient, and omnipotent. I'd still do wrong and unfair things. At best, when defined correctly, the three "omnis" are an effect of God's Nature.

Omnipresent: God is everywhere it is right and just for Him to be. Notice, Righteousness and Justice is the limitation. Is God inside the heart of an unbeliever? No.

Omniscient: God knows what is right and just for Him to know. He knows all the information that exists (the causes that are in place and the effects of those existing causes).

Omnipotent: God can do anything that is right and just for Him to do. Again, His power is limited by Righteousness and Justice. God can do anything He wants to things that don't have a mind/soul because they don't have a will, such as inanimate objects, weather, and animals.

I came across these "omnis" multiple times in my university studies. I was ready for it as I had already learned the correct perspective about the "omnis." What I found surprising was yet another support for why the three "omnis," as they're traditionally defined, cannot be the definition of God's Nature.

One of my university classes, a class that was specifically for my Theology degree, presented a fourth "omni": omnibenevolence (always loving). The professors instinctively knew the three "omnis" had contradictions and didn't account for how God loves, so they created a new omni in an attempt to fill the gap that the three "omnis" created, a gap that is really just a contradiction these professors tried filling with yet another contradictory term. Does God always love? Will it be an act of love when God declares judgment against the eternally damned?

To my lack of surprise, I've recently heard of yet another "omni" being embraced by Christians and theologians. It seems more people than I realized, at least unconsciously, that the three "omnis" are contradictory and continue to add more information in order to make up for the flaws in that model. The newest "omni" is omnitemporal. This "omni" states that God has reference to all time, including the ability to go ahead in time.

Lenhart stated above that "Determinism believes God sees every moment as occurring at the same time, which means the future has already happened and we don't have free will." Boethius stated the same thing.

We can see omnitemporal is unjust and therefore cannot be an aspect of God's Nature. When a model needs more information added to it in order for it to make sense, it's an effect of that model being flawed. When a model is non-contradictory, the model becomes tighter the more information you throw at it. For instance, when you understand the model for God as presented in the first edition (*Modeling God*), hearing someone teach the concept of omnitemporal only makes the non-contradictory model make more sense, and the result is it immediately shows the flaw in the "omnis" model.

The concept of omnitemporal would be yet another effect of God's Nature, not a cause. In fact, it becomes simple to see that omnitemporal would be an effect of omnipresent, which is an effect of God knowing all the information that exists, including possible future effects of causes that are in place. Knowing a possible future effect, because you know the causes in place, is not the same as going ahead in time.

The Bible and TULIP

John 3:16 is the ultimate verse that contradicts the five points of Calvinism:

> *For God so loved the world, that he gave his only begotten Son, that whosoever believeth in him should not perish, but have everlasting life.*

Total depravity: Could a totally depraved person make the decision to confess with his mouth that Jesus is Lord and believe in his heart that God raised Him from the dead? Wouldn't that be considered a good decision?

Unconditional election: "Whosoever believeth on him" implies a person can choose not to believe. This is a choice, not something God chose ahead of time.

Limited atonement: God loved the world and "whosoever" implies that the atonement was and is available to everyone; it is not limited.

Irresistible grace: If those elected cannot resist God's grace to save them, why does it require a person to choose to believe in Christ?

Preservation of the saints: Again, a point of Calvinism reliant on a person not having free will, John 3:16 shows the choice of belief or unbelief is up to the person. This choice, to believe or not believe, can happen at any time. This means a person can choose not to believe after they've chosen to believe.

However, it gets even deeper than that when you start to notice how the verses that Calvinists use to support one part of TULIP contradict other parts of TULIP.

Total depravity: Look at the following verses used to support total depravity. The point here that none is righteous, no, not one (including both the Jews and the Gentiles) would contradict Calvinism's unconditional election point that God elected believers. Are they believers? Or are they totally depraved?

> *What then? are we better than they? No, in no wise: for we have before proved both Jews and Gentiles, that they are all under sin; As it is written, There is none righteous, no, not one: There is none that understandeth, there is none that seeketh after God. They are all gone out of the way, they are together become unprofitable; there is none that doeth good, no, not one.* (Romans 3:9-12)

Unconditional election: The following verse is used to support the election of the saints, yet it contradicts the totally depraved point when it teaches that people would have fruit that would remain. Can a totally depraved person produce good fruit?

> *Ye have not chosen me, but I have chosen you, and ordained you, that ye should go and bring forth fruit, and that your fruit should remain: that whatsoever ye shall ask of the Father in my name, he may give it you.* (John 15:16)

Limited atonement: Not only does this verse directly contradict this very point (Esau being hated doesn't mean God didn't love him), it also contradicts the point that an elect cannot resist God's will to save them. Esau's life went the way it did because of the choices he made, not because of God enforcing His will upon Esau.

> *As it is written, Jacob have I loved, but Esau have I hated.* (Romans 9:13)

Irresistible grace: the following verses, used to support the irresistibility of God's grace, contradict the total depravity, the limited atonement, and the unconditional election points. Someone totally depraved wouldn't understand something taught to them of God, that would require some good. It also says that all will be taught of God, meaning everyone will be given a choice, not a limited elect.

> *No man can come to me, except the Father which hath sent me draw him: and I will raise him up at the last day. It is written in the prophets, And they shall be all taught of God. Every man therefore that hath heard, and hath learned of the Father, cometh unto me. Not that any man hath seen the Father, save he which is of God, he hath seen the Father.* (John 6:44-46)

Preservation of the saints: The "pluck them out of my hand" verse does not account for the person who chooses to be in God's hand; it just states that other men wouldn't be able to pluck them out of God's hand.

> *And I give unto them eternal life; and they shall never perish, neither shall any man pluck them out of my hand. My Father, which gave them me, is greater than all; and no man is able to pluck them out of my Father's hand.* (John 10:28-29)

At the core of the five points, we see that each of the points attempts to cover the flaws of the other four.

Glorify

Much of the Calvinist doctrine comes down to the term glory/glorify. Let's look at a couple of verses in Romans 1 to clear up what this actually means, and again, you'll notice that the very verses that the Calvinists use to support their doctrine actually contradict their own doctrine.

> *For the invisible things of him from the creation of the world are clearly seen, being understood by the things that are made,*

> *even his eternal power and Godhead; so that they are without excuse:* (Romans 1:20)

God is invisible, and the proof of God's existence is clearly seen through the physical things that God made. This was done so that everyone has the information to make the right choice. The God-given principle of causality proves the existence of a First Cause.

God created our brains in a way that we innately embrace this truth, whether we're aware of it or not. Trying to ignore causality would give the same effects as trying to damage your brain and damage your faith. Verse 20 states that everyone is without excuse if they don't have faith in the existence of God. However, this was only the first half of the reason why God's wrath is against non-believers.

> *Because that, when they knew God, they glorified him not as God, neither were thankful; but became vain in their imaginations, and their foolish heart was darkened.* (Romans 1:21)

The word glorified was G1392 "doxazo." The definition is "to think, praise, honor" and resulted in a person thinking well of the subject they glorified. The ultimate cause of people being non-believers is that these people choose not to think well of God; in fact, they don't even give God thanks!

The cause for people not glorifying God: vain (unprofitable) in their inward reasonings (imaginations) and their senseless (foolish; without understanding) heart was darkened (effect of a bad thought process).

God did His part, and the reason people do not believe is that they choose not to recognize the value within God. This happens even though they know a First Cause must exist, and working against causality results in their thought processes becoming worse.

The rest of Paul's letter to the Romans was based on this critical cause: God does not think well of (glorify) people who choose not to think well of (glorify) God, and the result is God's wrath is against these people. The principle of Justice guides God's response to the individual!

CHAPTER 5

Moses and Jeremiah

GOD'S WILL IS counterintuitive; otherwise, everyone would understand this. The only way to prove God's will is with the Bible. We are going to cover several passages that many people don't even realize are in the Bible, let alone see these passages relevant to God's will. We will take our time with these, and while it may seem like this is too detailed, it is important because this entire edition is based on the passages in this chapter and the next chapter. I like to see these chapters as filling in one of the images on the wall.

The first passage we will look at is Numbers 14:11-21. The context was set by the previous chapter (Numbers 13), which covered the story where God told Moses to send out twelve spies into Canaan, the land God promised them, and asked them to do the following five things:

1. See what the land is, whether it is good or bad, fat or lean.
2. See the people that dwell in the land, whether they are strong or weak, few or many.
3. See the cities the people dwell in, whether they are strongholds or camps.
4. See if there is wood in the land.
5. Bring back fruit from the land.

After forty days, the spies returned. While the spies did the five things God asked, ten of the twelve spies also added that they would not be able to go up against the inhabitants of the land because they were giants. However, one of the spies (Caleb) said they ought to go up at once and possess it because they were able to overcome the giant inhabitants.

Numbers 14 began with the children of Israel siding with the ten spies, so they spent the night weeping and then murmuring against Moses and Aaron. However, Joshua, one of the two spies who did not give an evil report, encouraged the congregation not to rebel against God because He was with them and would give them the victory. The people responded by wanting to stone Joshua, Caleb, Moses, and Aaron.

The Lord appeared in front of all Israel in the tent of meeting in order to speak with Moses.

> *And the Lord said unto Moses, How long will this people provoke me? and how long will it be ere they believe me, for all the signs which I have shewed among them? I will smite them with the pestilence, and disinherit them, and will make of thee a greater nation and mightier than they. (Numbers 14:11-12)*

God stated His will. He would smite them, disinherit them, and begin again by making a greater and mightier nation for Moses to lead. This is God's will.

> *And Moses said unto the Lord, then the Egyptians shall hear it, (for thou broughtest up this people in thy might from among them;) And they will tell it to the inhabitants of this land: for they have heard that thou Lord art among this people, that thou Lord art seen face to face, and that thy cloud standeth over them, and that thou goest before them, by day time in a pillar of a*

cloud, and in a pillar of fire by night. Now if thou shalt kill all this people as one man, then the nations which have heard the fame of thee will speak, saying, Because the Lord was not able to bring this people into the land which he sware unto them, therefore he hath slain them in the wilderness. (Numbers 14:13-16)

Moses confirmed God's will was to kill all the people, and he appealed to God by stating that the other nations would say the Lord killed them because He was not able to bring this nation into the land that He swore to them.

And now, I beseech thee, let the power of my Lord be great, according as thou hast spoken, saying, The Lord is longsuffering, and of great mercy, forgiving iniquity and transgression, and by no means clearing the guilty, visiting the iniquity of the fathers upon the children unto the third and fourth generation. Pardon, I beseech thee, the iniquity of this people according unto the greatness of thy mercy, and as thou hast forgiven this people, from Egypt even until now. (Numbers 14:17-19)

Moses expressed his will that God would not kill these people by appealing to God's personality. More than that, Moses was stating his will to God.

And the Lord said, I have pardoned according to thy word: But as truly as I live, all the earth shall be filled with the glory of the Lord. (Numbers 14:20-21)

God saying "according to thy word" meant God would do Moses' will instead of His own!

If Calvinism is true, then why did God change His Mind if the future was already determined and God can do whatever He wants?

God said He will smite them and disinherit them and make a greater and mightier nation out of Moses. Moses talked (prayed?) to God and changed His Mind. The result was that God said He will do according to Moses' word after stating He was going to do something different.

Remember, God has the ability to consider He is wrong, which would also be the same ability to consider something other than His will. God was not wrong. God told Moses His will, but He embraced Moses' will.

Most Calvinists would respond with the contradictory explanation of God's will really being, "He didn't want to do this," etc. This would actually make God out to be a liar. God saying He was going to do something that He knew He was never going to do would be God lying. Do you believe God is a liar?

For now, we can see Moses' plea as a prayer, and we know the four parts to prayer identified in *Modeling God* (recall the Four Rs of Prayer):

1. **R**ecognize (Identify) you are praying to God
2. **R**einforce (Build) your faith through understanding and/or experience
3. **R**eference justice
4. Make your **R**equest

Look again at what Moses stated:

> *And now, I beseech thee, let the power of my Lord be great, according as thou hast spoken, saying, The Lord is longsuffering, and of great mercy, forgiving iniquity and transgression, and by*

no means clearing the guilty, visiting the iniquity of the fathers upon the children unto the third and fourth generation. Pardon, I beseech thee, the iniquity of this people according unto the greatness of thy mercy, and as thou hast forgiven this people, from Egypt even until now. (Numbers 14:17-19)

Moses spoke to God about what God Himself had said about Himself. Then Moses made his request according to justice: how God had been forgiving the people from Egypt even until now.

For now, here are three questions to ponder:

1. Could anyone else have convinced God to change His Mind in this situation?
2. Could everyone have convinced God to change His Mind in this situation?
   (If God never really meant to do this, then anyone, and possibly everyone, could have convinced Him.)
3. Could Moses always convince God to spare people in any/every situation?

Jeremiah

The second passage covered in this chapter is Jeremiah 15:1-6. The context was set by the previous chapter (Jeremiah 14), which covered the story where God told Jeremiah He will remember the Israelites' iniquity and visit their sins.

God told Jeremiah not to pray for the Israelites' good because when they fast, He will not hear their cry, when they make burnt offerings He will not accept them, but consume them by the sword, the famine, and the pestilence. This is about 800 years after the previous passage with Moses, so God had decided that continuing to forgive would enable Israel. He was done forgiving.

Jeremiah told God that the other prophets were saying they wouldn't see sword nor famine. God responded by saying they were lying and those prophets would be consumed by sword and famine. In fact, from the beginning of the Book of Jeremiah, God tried every possible way to illustrate to Judah their level of wickedness with analogies using a girdle, marriage, and divorce.

Here was how Jeremiah 15 began:

> *Then said the Lord unto me, Though Moses and Samuel stood before me, yet my mind could not be toward this people: cast them out of my sight, and let them go forth.* (Jeremiah 15:1)

God said that neither Moses nor Samuel would be able to change His Mind about this decision to do His will. We have the answer to Question #3 from the previous section! "Could Moses always convince God to spare people in any/every situation?" No.

It also looks like we got an answer to Question #2. "Could everyone have convinced God to change His Mind in this situation?" No. If specific people couldn't convince God not to do His will in this case, there must be people who could not convince God to do His will in the previous case.

As for Question #1, "Could anyone else have convinced God to change His Mind in this situation?," this could be answered if we knew the measure for what it takes to be able to prevent God from doing His will.

It looks like we need to read more passages to determine the answer. Let's continue with this passage.

> *And it shall come to pass, if they say unto thee, Whither shall we go forth? then thou shalt tell them, Thus saith the Lord; Such as are for death, to death; and such as are for the sword, to the sword; and such as are for the famine, to the famine; and such as are for the captivity, to the captivity. And I will appoint over them four kinds, saith the Lord: the sword to slay, and the dogs to tear, and the fowls of the heaven, and the beasts of the earth, to devour and destroy. And I will cause them to be removed into all kingdoms of the earth, because of Manasseh the son of Hezekiah king of Judah, for that which he did in Jerusalem.* (Jeremiah 15:2-4)

God gave the reason why He would cause all of Israel to be put under the kingdoms of other nations: What Manasseh did in Jerusalem. That implied justice. It was right for God to justly respond to Israel by putting Israel in captivity.

> *For who shall have pity upon thee, O Jerusalem? or who shall bemoan thee? or who shall go aside to ask how thou doest? Thou has forsaken me, saith the Lord, that are gone backward: therefore will I stretch out my hand against thee, and destroy thee; I am weary with repenting.* (Jeremiah 15:5-6)

In the previous passage, we saw Moses tell God to have mercy and turn away from the evil the people deserved and not do His will. The word repent means to turn away. In this passage, it is more than 800 years after Moses prevented God from doing His will. In that time, God continued to have mercy when Israel's sin got worse to the point God got weary of repenting (turning away) from doing His will.

Relative to Calvinism, there are two questions this raises:

1. If God is "All-Powerful" then how is Israel able to make Him weary?
2. If God is able to go ahead in time, why was Israel able to make Him weary?

Clearly, rather than determinism, God is in the moment and can see the effects of all the causes that exist. What is happening?

For more than 800 years, God continued to show mercy by repenting from equaling out justice, however, He reached the point where allowing it to continue would have become enabling. It looks like this justice accumulated to the point that God had to do the worst thing in response.

Notice, God gave the reason for His judgment: Manasseh. One of the things I love about the Bible is that it shows God wants us to understand Him! In fact, God being unable to be understood by us would be an example of something God can't do: explain Himself to us. However, the Bible consistently stated we can understand God and, even more, God is constantly explaining Himself so we do understand Him!

Summary

Taking these two passages together begins the process of building a model to understand why God does and doesn't move for and against people! It shows what causes God to do His will and not do His will.

Why couldn't Moses and Samuel change God's Mind in this instance? Could anyone change God's Mind in this instance?

Ultimately, all of our questions concerning God's will can be answered if we understand the objective measure that determines whether a specific person is able to change God's Mind.

Dr. Joel Swokowski's Commentary

Adding to God's Word

The major mistake made by ten of the spies who were sent out to get a survey of the land was adding to God's word. They could have easily rationalized this by saying, "Well, we just thought you wanted all the information. We're just being contrastive." Unfortunately, the report the ten spies gave caused the Israelites' faith to lessen, preventing them from doing God's will. Even if their intentions were good, adding to God's word made the situation worse.

Eve did the same thing. Genesis 2:16-17 recorded that Adam and Eve were not to eat of the fruit, but Eve, when questioned by the serpent, added the part about not touching the fruit. This was the first example of a human adding to God's word. This was the first example of man-made religion: a man-made list of rules to follow for a system of belief.

I covered how the religious authorities during Jesus' time did the same thing. They added rules, regulations, and laws onto the Torah (God's word) in an effort to make it easier to follow God's word. Notice, even if their intentions were good, it resulted in a controlling government that murdered Jesus.

Jesus confronted those religious authorities time and time again, and as we saw in an earlier chapter, Jesus was, at one point, as specific as stating, "Making the word of God of none effect through your tradition, which ye have delivered: and many such like things do ye" (Mark 7:13).

We can't make God's word better, no matter how good our intentions are! When we add to God's word, we make it worse, and we end up hurting people. This is what Calvinists are doing.

CHAPTER 6

Ezekiel and Hezekiah

AS WE CONTINUE to dig deeper into the Bible to understand God's will, the third passage that will bring more clarity is Ezekiel 14:12-23, which took place less than 40 years after the previous passage from Jeremiah. The ten tribes of Israel had been in captivity for over 100 years, while the rest of Judah was about to go into captivity with the fall of Jerusalem.

The context was set by the previous chapter (Ezekiel 13), which covered the story where God told Ezekiel to prophesy against the prophets who prophesied out of their own hearts, which was reminiscent of what was happening in the previous passage with Jeremiah. The false prophets were stating that there would be peace in Judah and Jerusalem wouldn't fall, with the final two tribes of Israel avoiding going into captivity in opposition to what God had previously stated.

Ezekiel chapter 14 began with elders sitting in front of Ezekiel trying to inquire of God through him and God stating that people who had taken idols into their heart and put the stumbling block of iniquity before their face would be answered directly by God. God encouraged them to turn from their idols; otherwise, God would set His face against that person and make an astonishment of them.

Here is how Ezekiel 14 continues:

> *The word of the Lord came again to me, saying, Son of man, when the land sinneth against me by trespassing grievously, then will I stretch out mine hand upon it, and will break the staff of the bread thereof, and will send famine upon it, and will cut off man and beast from it: Though these three men, Noah, Daniel, and Job, were in it, they should deliver but their own souls by their righteousness, saith the Lord God.* (Ezekiel 14:12-14)

This all began with the land sinning against God, and that allowed God to justly send a famine. Remember, Jeremiah also mentioned famine as a way God would judge. The difference was God referenced three men: Noah, Daniel, and Job. Notice, unlike the previous two passages, God wasn't saying these three men could change His Mind/will. Instead, He stated only these three men would be able to deliver their souls from the judgment.

However, there was a more specific reason as to the cause of these three men delivering their own souls from judgment: by their righteousness! Didn't Calvinism state that man is totally depraved?

Anyone who read *Modeling God* realizes this verse is speaking of reward. There is a Righteousness that leads to salvation, and we realized that is God's Righteousness through grace. Jesus' death on the cross created infinite value to cover the sins we do against God.

On the other hand, all of us can do righteous works and those result in spiritual reward. Some of these are the works God predestined for us to walk into. Since it is our choice to do it, we get the credit for it. This spiritual reward is completely different from the value Jesus generated leading to salvation. Let's keep going.

> *If I cause noisome beasts to pass through the land, and they spoil it, so that it be desolate, that no man may pass through because of the beasts: Though these three men were in it, as I live, saith the Lord God, they shall deliver neither their sons nor daughters; they only shall be delivered, but the land shall be desolate.* (Ezekiel 14:15-16)

"Noisome beast" is in addition to the three judgments Jeremiah mentioned. The key point is these three men wouldn't be able to deliver sons or daughters. Noah did deliver his sons during God's previous judgment of the flood. Why wouldn't he be able to do it here?

> *Or if I bring a sword upon that land, and say, Sword, go through the land; so that I cut off man and beast from it: Though these three men were in it, as I live, saith the Lord God, they shall deliver neither sons nor daughters, but they only shall be delivered themselves.* (Ezekiel 14:17-18)

Jeremiah mentioned the sword as a way God would judge. Notice, God reemphasized that sons and daughters wouldn't be saved.

> *Or if I send a pestilence into that land, and pour out my fury upon it in blood, to cut off from it man and beast: Though Noah, Daniel, and Job were in it, as I live, saith the Lord God, they shall deliver neither son nor daughter; they shall but deliver their own souls by their righteousness.* (Ezekiel 14:19-20)

Jeremiah mentioned pestilence as a way God would judge. Again, sons and daughters wouldn't be saved and God emphasized the cause was by their righteousness. Ezekiel has mentioned four possible judgments.

> *For thus saith the Lord God; How much more when I send my four sore judgments upon Jerusalem, the sword, and the famine, and the noisome beast, and the pestilence, to cut off from it man and beast? Yet, behold, therein shall be left a remnant that shall be brought forth, both sons and daughters: behold, they shall come forth unto you, and ye shall see their way and their doings: and ye shall be comforted concerning the evil that I have brought upon Jerusalem, even concerning all that I have brought upon it. And they shall comfort you, when ye see their ways and their doings: and ye shall know that I have not done without cause all that I have done in it, saith the Lord God.* (Ezekiel 14:21-23)

God also said that there will be others who are themselves saved, and seeing their ways and doings will be a comfort to Ezekiel because he would know that God did not do this without cause. What was the cause? How did God measure all of this?

Let's summarize all three passages before the thrilling conclusion.

In the first passage, God said it was His will to destroy the people and to make a better and greater nation out of Moses. Moses opted for saving the people and God followed Moses' will!

In the second passage, God said that Moses and Samuel couldn't convince Him to save the people. Both of these passages had God looking at the people as a group when He expressed His will.

In this third section, God said that Noah, Daniel, and Job would only be able to save themselves and not even their own sons and daughters, and there was a remnant that was saved all due to the righteousness of their ways and doings. In fact, God ended the last passage by saying all of this would be a comfort to Ezekiel because Ezekiel would know the judgment didn't come without cause! Even though God had a will for the group of people, He had a way individuals could have a will different from the group's will.

Here are some points to ponder:

- If it was all pre-determined, as the Calvinists purport, God didn't have to prove or explain anything to anyone because the cause was that He wanted it.
- The concluding explanation was that everything happened because of the people, as an effect of their decisions!
- Remember, salvation occurs with God's Righteousness through grace. Reward occurs with our righteousness (works). Works (our righteousness) do not save us (salvation). Five-point Calvinism (the man-made traditional explanation for God's will) says there is nothing good in man; he is totally depraved. This passage said God measured value affecting His will in response to the person's righteousness. It looks as if God's will is according to the same currency as reward rather than salvation and that God doesn't mix them together.

Questions

1. What really drove the fact that God now had a greater and better will for Moses? (Reread the previous chapter, then reread the first Bible passage of this chapter! Moses qualified for a greater will because when he was abused by the people, he handled that abuse well. This resulted in God wanting to begin again with only Moses!)

2. How was Moses able to save the people in the desert when he (and Samuel and Noah and Daniel and Job) couldn't in Jerusalem?
3. How was Noah able to save his sons (and daughters-in-law) before the flood, but not be able to save them in Jerusalem?
4. If God is ahead in time and it is all pre-determined, why is God contemplating the "ifs"?

Ultimately, we were trying to answer this question:

> What was the objective measure that determined whether a specific person is able to change God's Mind?
>
> Was it reward (spiritual value from righteousness)?

Hezekiah

The fourth and final passage is II Kings 20:1-6, which predates the passages from Jeremiah and Ezekiel. In order to set up the passage we are focused on, let's look at II Kings 18:1-7:

> *Now it came to pass in the third year of Hoshea son of Elah king of Israel, that Hezekiah the son of Ahaz king of Judah began to reign. Twenty and five years old was he when he began to reign; and he reigned twenty and nine years in Jerusalem. His mother's name also was Abi, the daughter of Zachariah. And he did that which was right in the sight of the Lord, according to all that David his father did. He removed the high places, and brake the images, and cut down the groves, and brake in pieces the brasen serpent that Moses had made: for unto those days the children of Israel did burn incense to it: and he called*

> it Nehushtan. He trusted in the Lord God of Israel; so that after him was none like him among all the kings of Judah, nor any that were before him. For he clave to the Lord, and departed not from following him, but kept his commandments, which the Lord commanded Moses. And the Lord was with him; and he prospered whithersoever he went forth: and he rebelled against the king of Assyria, and served him not.

Hezekiah was a righteous king who was completely in line with God to the point it was stated he was the greatest ever of all the kings of Judah. The rest of II Kings 18 explained how in the sixth year of Hezekiah, Samaria was taken, and Israel was carried away into captivity in Assyria because they didn't obey the voice of God even in all that Moses commanded.

The king of Assyria came against Judah, and in II Kings 19, Hezekiah went into the temple to pray to God. This prayer was covered in *Modeling God* as an example of how all the prayers in the Bible followed a pattern consisting of four parts (the four R's of prayer), which we reviewed in the previous chapter. Isaiah came to Hezekiah to say God heard him, and the result was that the angel of Jehovah smote 185,000 Assyrian soldiers in one night.

Here was how the next chapter, II Kings 20 begins:

> In those days was Hezekiah sick unto death. And the prophet Isaiah the son of Amoz came to him, and said unto him, Thus saith the Lord, Set thine house in order; for thou shalt die, and not live. (II Kings 20:1)

God told Hezekiah he was going to die. His death was so close, God told Hezekiah to put his house in order, which is God's will. Remember, this illness would end in death.

> *Then he turned his face to the wall, and prayed unto the Lord, saying, I beseech thee, O Lord, remember now how I have walked before thee in truth and with a perfect heart, and have done that which is good in thy sight. And Hezekiah wept sore.* (II Kings 20:2-3)

Hezekiah made a request to God not to die by recognizing God and referencing justice through stating his righteousness.

> *And it came to pass, afore Isaiah was gone out into the middle court, that the word of the Lord came to him, saying, Turn again, and tell Hezekiah the captain of my people, Thus saith the Lord, the God of David thy father, I have heard thy prayer, I have seen thy tears: behold, I will heal thee: on the third day thou shalt go up unto the house of the Lord. And I will add unto thy days fifteen years; and I will deliver thee and this city out of the hand of the king of Assyria; and I will defend this city for mine own sake, and for my servant David's sake.* (II Kings 20:4-6)

God responded before Isaiah had left the middle court! God even said the reason He was healing Hezekiah was because of Hezekiah's prayer. What was actually going on from the perspective of God's will?

God's will: Die now and keep your spiritual value (reward stored in heaven; Matthew 6:20).

Hezekiah's will: Have less spiritual value because he spent it on the physical in order to die at a later time.

Hezekiah changed God's will with prayer according to spiritual value from his righteousness.

God said He would *add* fifteen more years to Hezekiah's life. Whenever Hezekiah died, God had intended that Hezekiah die fifteen years *before* that time. As it turned out, Hezekiah died fifteen years later.

Questions

1. If God can go ahead in time, why was He wrong about Hezekiah dying? ("thou shalt die, and not live")
2. If God knew Hezekiah wasn't going to die, but said it because (insert your reason here), wasn't this lying? (Remember, God cannot lie. It is impossible. Stating God lied could be seen as blasphemy.)
3. How can God see the future and be wrong or need to lie? Isn't being able to cheat and still being wrong the opposite of a God who can be seen as having excellent ability?
4. Doesn't a belief in Calvinism's definition of predestination result in God being a liar, wrong, or not "all-powerful" according to the Bible?
5. Why did God say Hezekiah was going to die?
6. Why did Hezekiah have to pray in order not to die?
7. What was the final result of Hezekiah living fifteen more years?

You have all the information you need to answer these questions in a non-contradictory and non-blasphemous manner if you look at this entire chapter and the previous chapter.

Please take a moment to answer all of these before continuing to the next chapter.

Dr. Joel Swokowski's Commentary

Impossible for God to Lie

In the Bible, there were very few things that were impossible. There were a lot of things that were possible, but weren't going to happen. If we take the things in the Bible that were possible, but weren't going to happen and make them impossible, then we have created the man-made doctrine of predestination, and we would result in God being unjust.

In order to understand this at the depth we need to, let's review the God-given principle of causality.

When we speak about things "happening," we are really discussing an effect. The happening is an effect. However, causality says that effects are the natural occurrence of a cause. Remember, causality is both cause and effect; they are halves of an inseparable whole. So, discussing anything happening requires us to identify a cause or even multiple causes.

Said another way, the ability for an event to happen depends on the ability for the cause(s) to exist. Allow me to elaborate: if a cause doesn't exist or is not possible, then the effect is impossible.

So, in order to prove something is impossible, you would have to know the causes that would result in the impossible action. It is not good enough to say something is impossible just because you don't know the causes.

For example, the number of experts (scientists, engineers, etc.) that positively stated that heavier-than-air flight was impossible before the Wright brothers' flight is too many to count, let alone reference.

This method of transportation has become such a part of the culture in which we live that we no longer question its impossibility or even improbability. The causes of heavier-than-air flight were not yet known to those experts who stated its impossibility, but we know today it was not impossible.

If the cause does exist and is possible, but the odds of it happening are very, very slim, then the effect is not impossible, no matter how improbable. A simple example of this is when you shuffle a deck of cards. You can take a standard 52-card deck and shuffle it wherever you are right now, and the probability of a card sequence in the exact order that you come to is 1 in 80658175170943878571660636856403766975289505440883277824000000000000.

I don't know what to call that number other than to write it out and show you how ridiculously slim those odds are. Despite how low the probability was, that sequence did happen. What's even more mind-boggling is that, although even more improbable, it is entirely possible to come up with that same sequence again. Although that may never happen, it is possible, no matter how unlikely.

So, when we read the word impossible, it means the cause does not exist and/or is not possible to exist.

Let's get to the passage of scripture for our example:

> *That by two immutable things, in which it was impossible for God to lie, we might have a strong consolation, who have fled for refuge to lay hold upon the hope set before us:* (Hebrews 6:18)

The Bible said, "it was impossible for God to lie."

What is lying?

Lying is an intentional misstatement of reality.
Lying is not a fact.
Lying is not truth.
Lying does not create in the long term.

Even an unintentional misstatement about reality is wrong, and we're saying here that lying is an intentional misstatement about reality: still wrong.

The point of Modeling God is to prove that God is always and completely Righteous and always and completely Just. This is God's Nature, and due to the "always and completely" qualifier, we know that God cannot act apart from this Nature. Said another way, it is impossible for God to be wrong or unjust.

If lying is wrong, and God doesn't have the causes within Him to be wrong, then it is impossible for God to lie.

What cause would God have within Him that would result in Him lying? There would have to be some "darkness" within Him, and we have seen from I John 1:5 *that God is light, and in him is no darkness at all.* A huge implication here: Calvinists believe God has darkness within Him. Isn't this blasphemous?

Hezekiah

If you doubt the explanation given about Hezekiah, this story was also told from Isaiah's perspective in Isaiah chapter 38. In that chapter, Isaiah even documented a song that Hezekiah wrote that gave further proof that God added fifteen years to Hezekiah's life. Here are some highlights from Isaiah 38, from Hezekiah's perspective:

> *I said in the cutting off of my days, I shall go to the gates of the grave: I am deprived of the residue of my years.* (Isaiah 38:10)

> *O Lord, by these things men live, and in all these things is the life of my spirit: so wilt thou recover me, and make me to live. Behold, for peace I had great bitterness: but thou hast in love to my soul delivered it from the pit of corruption: for thou hast cast all my sins behind thy back.* (Isaiah 38:16-17)

God delivered Hezekiah from the death coming towards him. God gave Hezekiah extra time in response to Hezekiah's righteousness and Hezekiah's request.

CHAPTER 7

God's Will Explanation

LET'S LOOK AT a summary of the previous Bible passages:

- Moses was able to change God's Mind, changing God's will relative to a group of people.
- Later, God said even if Moses and Samuel were present, they could not change His Mind; therefore, they were not changing God's will relative to a group of people.
- God said that if Noah, Job, and Daniel were present, they would only be able to save themselves from judgment by their own righteousness and not their sons and daughters; there was a remnant that was likewise saved due to the righteousness of their ways and doings. In fact, God ended that passage by saying all of this would be a comfort to Ezekiel because Ezekiel would know the judgment didn't come without cause! Even though God had a will for the group of people, He had a way individuals could have a different will according to God.
- Hezekiah changed God's will toward him (individual) with prayer according to spiritual value from his personal righteousness.

While all four passages disprove Calvinism, the implications of Hezekiah's story essentially remove determinism as an explanation for God's will because of determinism's catastrophic conclusions about God.

The future has not already happened; otherwise, God is wrong and/or a liar. People who hold to this view, regardless of the long and complicated explanation they give, are stating their will that they think God is wrong and/or a liar. The Calvinist's definition of predestination fails and needs to be discarded.

The only non-contradictory explanation I know of that is consistent with the entire Bible is that God has a will for each of us that changes depending on our response. This would make it impossible for God to be wrong because all the responsibility is on us.

Notice the word is impossible, meaning "not possible." Some of the explanations from Calvinists say it is possible, but God makes sure He isn't wrong and/or unjust. The Bible tells us there is a higher standard: impossible.

We ended the previous chapter asking seven questions relative to the Hezekiah story and encouraged you to take a moment to answer all seven questions before continuing to the following hard explanation:

God's will for Hezekiah was that he would die. God wants what is best for us in the long term. Hezekiah's death would have been the best thing for him in the long term. Hezekiah was going to heaven. Because Hezekiah was righteous, he had a lot of reward and dying at that time would have ensured Hezekiah would go to heaven with a lot of reward. Hezekiah avoided dying at that time through prayer. Remember, part of prayer is the expression of the will of the individual to turn spiritual value into physical value according to justice. (See *Modeling God* for the entire explanation of prayer.) So, now Hezekiah would go to heaven with less reward. However, that isn't the entire story.

When it comes to the first three passages, Hezekiah (much like Moses) chose to spend his value in a way different from what God wanted. God wanted Moses to spend it on being a better and stronger nation. Jeremiah was not in the position to state his will in the situation. God was saying that even if Moses and Samuel were in the position to state their will, they wouldn't have enough value to change God's Mind, whereas Moses had enough in the previous situation. As for the situation in Ezekiel, God shared more detail about spiritual value to say these men would only have enough to save themselves from the impending judgments, and not enough to save their kids.

The seventh and final question from the previous chapter was: "What was the final result of Hezekiah living fifteen more years?" Remember, Moses expressed his will and it had an effect on the nation. What about Hezekiah?

> *And Hezekiah slept with his fathers: and Manasseh his son reigned in his stead. (II Kings 20:21)*

Hezekiah died at the end of the chapter. The next chapter begins like this:

> *Manasseh was twelve years old when he began to reign, and reigned fifty and five years in Jerusalem. And his mother's name was Hephzibah. And he did that which was evil in the sight of the Lord, after the abominations of the heathen, whom the Lord cast out before the children of Israel. (II Kings 21:1-2)*

Notice, Hezekiah's son was twelve years old when Hezekiah died. Manasseh would not have been born if Hezekiah had died before his prayer. Remember this verse from the Jeremiah passage?

> *And I will cause them to be removed into all kingdoms of the earth, because of Manasseh the son of Hezekiah king of Judah, for that which he did in Jerusalem.* (Jeremiah 15:4)

In the Jeremiah passage, God said that He would cause (through justice) Judah to go into captivity because of Manasseh, which would put all twelve tribes in captivity.

This meant the other part of the benefit of Hezekiah's death was to Judah (a group will) so that Manasseh would not be born. God was not the cause of Judah going into captivity. Manasseh was the cause.

Now we see, God saw Hezekiah only had enough value to live fifteen more years, and that wouldn't be enough to train up a righteous king, so God's will for Hezekiah *and* Judah was to prevent Manasseh's birth by having Hezekiah die. Hezekiah didn't have a son to take over the kingdom, so knowing he only had fifteen years to live would cause him to focus on creating an heir.

God tried to influence Hezekiah to make the right decision: don't have an heir to the throne because you wouldn't be able to ensure he was mature enough to rule in a righteous manner. Does it really take a genius to think that a twelve year old was not going to be wise and show restraint when he was put in complete command of the kingdom? God was not at fault, either through causing the captivity, nor failing to try to stop the captivity.

It was God's plan for Israel to continue to grow in faith in God. It was God's will *for* Hezekiah to die. How do you feel about this?

Our traditional view of short-term comfort being equal to God's will keeps us from seeing the non-contradictory explanation for these passages. Worse, it causes people to come up with contradictory fairy tales like "predestination" that are actually blasphemous and make the word of God of none effect.

Every time God moved for, moved against, or didn't move for someone, it was according to the principle of Justice. I've heard Calvinists ask why God moved against certain people in the Bible and didn't move against others, only to conclude it was essentially causeless: "God knows, and we aren't supposed to know."

The reason is everyone God moved for or didn't move against did righteous works that put them on the right (positive) side of Justice. God specifically stated He wanted us to know this happens for a cause. If these Calvinists really wanted to know the answer, they would confess and repent. If their goal is to appear to be right, they won't be able or willing to understand this explanation.

Remember, this has nothing to do with salvation. Salvation and rewards are two different models. Combining them means that you are making it possible to earn your salvation through works. You can only get salvation through God working through you (grace). God initiates the Righteous actions that result in salvation. However, you can get rewards from righteous actions you initiate. These righteous acts are like filthy rags compared to the Righteousness that leads to salvation. However, it looks like these righteous acts are associated with God's will.

Moses, Samuel, Noah, Daniel, and Job all did righteous acts while being surrounded by people doing acts of unrighteousness. In fact, Noah, Daniel, and Job being used as the best examples show they are the three people who accumulated the most spiritual value by the time Ezekiel lived. While we will see in later chapters more specifics about each of these people, here are some reasons why these men had so much value.

Noah was surrounded by an entire world of unrighteousness and he handled it well. Daniel repeatedly handled unjust personal attacks and refused physical wealth, while praying a prayer in God's place that led to what I believe is the greatest prophecy in the Bible, one for which Daniel continues to receive value for, but we will talk about this later in this edition. Job unjustly lost all his possessions and positively handled an experience that was mentally, emotionally, spiritually, and physically grueling.

It looks as if God was saying that it didn't matter how much value each of these three men personally had; it wouldn't be enough to cover the unrighteous works done by the entire group.

Summary

Moses, Samuel, Noah, Daniel, and Job all had value to change God's Mind or avert His judgment in their situations, but not necessarily enough for other situations. I'm not aware of another non-contradictory explanation, so believing something else is a statement of the will of the individual to justify themselves instead of God if they don't have a different explanation that completely lacks contradiction.

This summary leads to the very uncomfortable conclusion that not only can we hinder God's will, but it looks like God needs us to bring about His will! God has a plan that involves us, and how He brings about this plan also involves us.

This uncomfortable non-contradictory conclusion is what Calvinists try to avoid, and the only way Calvinists can do it is to say God can unilaterally bring about His will without any interaction from people; however, we have seen this leads to several contradictions as well as the conclusion that God is wrong and/or unjust and/or a liar. It also begs a non-contradictory explanation for the following from Jesus' mouth in Matthew 6:9-10:

> *After this manner therefore pray ye: Our Father who art in heaven, Hallowed be thy name. Thy kingdom come. Thy will be done in earth, as it is in heaven.*

Jesus told us to pray for God's will to be done on earth in the same way God is able to do His will in heaven. There is a place where God unilaterally rules: heaven! There is a place where God does not unilaterally rule: earth.

If God's will can be changed by people with spiritual value, this means God's will can be opposed by people with spiritual value, and that leads to the very uncomfortable summary of everything we have covered so far:

> God's will is done justly through people who have enough spiritual value (righteousness).

It looks like it is getting more uncomfortable as we continue to move further away from the wall. However, we ought to begin to be able to see an image on the left side of the wall of God working through a human resulting in paradise.

Dr. Joel Swokowski's Commentary

Lenhart spent Part I of this book explaining the right answer to God's will and God's plan. The biggest implication: this explanation proved Calvinism wrong.

This is a huge deal, not in proving Calvinism wrong, but in the method it was done. Most people (Christians) spend their time poking holes in belief systems, focusing on proving something wrong without giving the right answer. Lenhart did the opposite.

Giving a non-contradictory answer to any concept, including God's will, proves every other contradictory option wrong as an effect. This is why Calvinism has continued to exist, not because it is right and lacks contradiction, but because no one has given an answer without contradiction. Calvinists know they have contradictions in their belief. Their response tends to be either pointing out the contradictions in other explanations or stating we can't know the non-contradictory explanation.

It wasn't Lenhart's intent to prove Calvinism wrong, it was just an effect of teaching the only non-contradictory answer we are aware of for God's will that is based in truth. Let me be clear, it's not a sin to say a belief system is wrong, but it can be if you don't also provide a non-contradictory answer; otherwise, you don't know for sure the belief system is wrong.

One more point that bears repeating relative to *Modeling God*, Lenhart used Old Testament passages to determine the non-contradictory explanation for God's will. It took four passages and a lot of explanation to reach the conclusion presented in this chapter. Lenhart could have done the same thing in *Modeling God* and an edition of *Modeling God* that only used Old Testament passages would have been three to four times larger. The same could be said for an edition of *Modeling God* using other religious texts. The reason *Modeling God* was shorter is that Lenhart took advantage of New Testament passages from Paul and Jesus, both of whom essentially did the work of summarizing multiple passages from the Old Testament.

The More Powerful God

It sounds great to say God can do anything without limitation. I understand that people want to *feel* great about the God they worship, and what *feels* better than a God that is all-powerful and in complete control? Yet, this is based on a feeling and results in contradictions. I want to feel great about the God I worship, just not at the expense of worshiping the wrong God!

We tend to put the characteristics that we want the most onto God. We (humans) want to be all-powerful, all-knowing, and all-present. We (humans) want to time travel so we can know the future down to every minute detail. Yet, we (humans) tend to want these things for reasons that benefit us individually and end up destroying us in the long term.

Let's think about this for a moment and contrast two versions of "God" by asking: "Which God do you believe is more powerful?"

1. The God that knows everything that will ever happen (down to the minute detail), can do anything (without limitation), and is in complete control?
2. The God that makes every decision based on Righteousness and Justice (never wrong or unfair) and has to account for and adjust to every decision that every human being makes at every given moment?

Number 2 sounds like a description that only God could do. Number 1 sounds like a contradictory fantasy.

This reminds me of something else I often hear about God, especially from Christians and especially in response to something "bad" that has happened: the phrase "God allowed it." What does this mean? Did God have the ability to stop the "bad" thing but chose not to? Why or why not?

I think this is an attempt to reconcile free will with God being "all-powerful." There's a contradiction that people cannot resolve. They want to keep their free will and their emotional perspective about God being "all-powerful," and the only way to do that when, for example, a loved one dies unexpectedly is to say, "God allowed it to happen." This platitude may make you feel good in the moment, but it doesn't actually answer the questions about *how* and/or *why* these things happen…or *how* and/or *why* God did or didn't get involved.

I hope you can see just how important it is to understand God's Nature. Understanding and believing that God is always completely Righteous and always completely Just is the key to us (humans) always being able to show God is right in *what* He does, *why* He does it, and *how* He does it.

- When something doesn't make sense, I can respond by saying, "I may not understand, but I know God is right and just!"
- When something "bad" happens to me, I can respond by saying, "I don't know why this happened, but I know God is right and just!"
- When I'm interpreting the scriptures, I can look at every story through the lens, "No matter what I think this means, I know it cannot contradict God being right and just!"

Remember, this is about reward, not salvation. This is about how much you can help facilitate God's will, not whether you are going to the lake of fire or the new Jerusalem. When Jesus spoke of removing your eye so the body can be saved (Matthew 5:29), He was speaking of salvation. As it relates to this chapter, we can say, "Going without an eye in order to be saved probably means you are limited in how much you can facilitate God's will."

Before we begin Part Two, take a moment to realize half of *Modeling God* (Book 1) deals with salvation. In contrast, half of *Modeling God* (Book 2) and all of *Modeling God's Wills* (Book 1 and 2) deal with reward and sanctification. Do you see "Christians" spending three times more of their focus on reward and sanctification than the time they focus on salvation? Do you believe God sees salvation as a hurdle or a driver? Do you believe God sees sanctification as a hurdle or a driver? Which one are you more focused on? Why?"

PART TWO

Characteristics of God's Will

INSIDE

Chapter 8: Individual Will 121

Chapter 9: Group Wills 133

Chapter 10: Three Levels in God's Will 143

Chapter 11: Council Meetings 156

Chapter 12: Dissolve 171

Chapter 13: God's Four Measures for Judgment 184

Chapter 14: Prophecy 198

Chapter 15: How God Brings About His Will 209

Afterword 225

CHAPTER 8

Individual Will

WE CONCLUDED PART I by seeing that when it comes to the Bible, there are two different wills: individual and group. Let's look at God's individual will more closely.

We have seen salvation is based on Righteousness, whether the individual chooses through faith to allow God's Righteousness to flow through them (grace). On the other hand, we have seen that reward is based on justice, the person accumulates reward (spiritual value) through justice based on the righteousness of their interactions with others.

Salvation and reward are tied together when it comes to the person achieving perfection: the maximum reward is achieved by following grace. However, salvation and reward are two completely separate concepts. I believe Matthew chapter 6 is the best place to see the difference between these two concepts. Let's take a quick look at the beginning of that chapter to better understand reward.

> *Take heed that ye do not your alms (righteousness) before men, to be seen of them: otherwise ye have no reward of your Father which is in heaven. (Matthew 6:1)*

In this verse, Jesus used the word reward. This word cannot mean salvation; otherwise, people would lose their salvation by giving alms or doing a righteous work and having someone see the righteous work. Notice also that Jesus spoke of *your* righteousness, which sounds a lot like how Noah, Daniel, and Job would be able to avoid judgment.

In order to explain this concept to people, I talk about value (or spiritual value). I could just as easily use the word reward, but I've found that people think of reward as only a benefit. The definition of reward (G3408 "misthos") in this verse applies "in both senses, reward and punishment." Another way to say it is "to recompense both good and bad." It is based on justice. It evens things out. It is possible to be rewarded evil. *Modeling God* showed this reward or value is spiritual and can be accessed through prayer, as we saw with Hezekiah.

This first verse of Matthew 6 showed there is a positive reward that is either obtained here and now, or later. The positive reward is deserved because of an action and/or attitude. However, this means the ability to retain the positive reward could be lost because of an action or attitude that either gains an immediate positive reward from others in the moment or deserves a negative reward. This has nothing to do with salvation. You cannot earn your salvation. You can earn reward, and Jesus taught this.

Here is the same explanation as the previous paragraph using the word value:

> This first verse of Matthew 6 showed there is a value that is either obtained here and now, or later. The value is deserved because the individual gave a value in action and/or attitude. However, this means the ability to retain the value could be lost because the individual is using the value now through an

action or attitude that is itself a value or because of an action or attitude that requires the loss of a value through justice.

For example, in verse 1, the individual deserves a reward for giving alms. However, the person could get their spiritual value now because they are seen by men. It is not necessarily bad to be seen by men. It is a reward or spiritual value to be seen by men, so there is no reward or spiritual value left to gain in the long term.

> *Therefore when thou doest thine alms, do not sound a trumpet before thee, as the hypocrites do in the synagogues and in the streets, that they may have glory of men. Verily I say unto you, They have their reward.* (Matthew 6:2)

This verse is a more specific explanation of verse 1 because it has an example. This example said the reason these hypocrites gave was so that they may have glory from men, that is, men will think well of them. Jesus just showed an example of an action or attitude that was wrong. Now, He's going to show a right way to do verse 1.

> *But when thou doest alms, let not thy left hand know what thy right hand doeth: That thine alms may be in secret: and thy Father which seeth in secret himself shall reward thee openly.* (Matthew 6:3-4)

Notice Jesus gave a reason and a value. Also, notice the *how* was vital to this discussion. No one would disagree about the *what*: give alms. However, giving alms and drawing attention to yourself is Right-Wrong. Giving alms and not drawing attention to yourself is Right-Right. If we couldn't know the right *how* or *why* or if it isn't important, then why did Jesus focus on telling us? As we read more of Matthew 6, notice the *how* and *why* were the focus of Jesus' message.

> *And when thou prayest, thou shalt not be as the hypocrites are: for they love to pray standing in the synagogues and in the corners of the streets, that they may be seen of men. Verily I say unto you, They have their reward.* (Matthew 6:5)

Now the discussion has moved to prayer! Prayer is a complex doctrine. As we saw with Hezekiah, it is where our will intersects with God's will. We saw in *Modeling God* that prayer is based on spiritual value. Notice, praying to be seen results in you getting your reward now. This looks as if the prayer doesn't get answered because the value was received in a different manner. It looks like the *how* and the *why* negated the benefit of the *what*.

> *But thou, when thou prayest, enter into thy closet, and when thou hast shut thy door, pray to thy Father which is in secret; and thy Father which seeth in secret shall reward thee openly. But when ye pray, use not vain repetitions, as the heathen do: for they think that they shall be heard for their much speaking.* (Matthew 6:6-7)

This looks as if you get your prayers answered (reward) because the only way the value can come to you is in response to your prayer because you didn't get any value from others like the hypocrites did when praying in public. The discussion then moved to vain (not profitable) repetitions. Again, vain is a quantitative word. Jesus said that saying the same valueless things over and over does not result in getting our prayers answered.

> *Be not ye therefore like unto them: for your Father knoweth what things ye have need of, before ye ask him.* (Matthew 6:8)

Jesus said that God knows what we need before we ask. So, the point so far is that saying the same thing over and over in public is not going to get our prayers answered. Also, since God knows what we need, was Jesus saying we shouldn't pray?

> *After this manner therefore pray ye: Our Father which art in heaven, Hallowed be thy name. Thy kingdom come, Thy will be done in earth, as it is in heaven. Give us this day our daily bread. And forgive us our debts, as we forgive our debtors. And lead us not into temptation, but deliver us from evil: For thine is the kingdom, and the power, and the glory, for ever. Amen. For if ye forgive men their trespasses, your heavenly Father will also forgive you: But if ye forgive not men their trespasses, neither will your Father forgive your trespasses.* (Matthew 6:9-15)

Jesus said we ought to pray even though God already knows what we need. We ought to state our will. This prayer is also known as "The Lord's Prayer," and we referenced this at the end of the last chapter and the implications of Jesus telling us to pray for God's will to be done on earth as it is in heaven.

> *Moreover when ye fast, be not, as the hypocrites, of a sad countenance: for they disfigure their faces, that they may appear unto men to fast. Verily I say unto you, They have their reward.* (Matthew 6:16)

Next, Jesus transitioned to fasting. Again, the example regarded a hypocrite who got their reward now because of the *how* and the *why*.

> *But thou, when thou fastest, anoint thine head, and wash thy face; That thou appear not unto men to fast, but unto thy Father*

> *which is in secret: and thy Father, which seeth in secret, shall reward thee openly.* (Matthew 6:17-18)

Again, the example spoke about not getting rewarded now. Jesus had given several examples that followed the same pattern. It was time for Jesus to summarize His point.

> *Lay not up for yourselves treasures upon earth, where moth and rust doth corrupt, and where thieves break through and steal: But lay up for yourselves treasures in heaven, where neither moth nor rust doth corrupt, and where thieves do not break through nor steal: For where your treasure is, there will your heart be also.* (Matthew 6:19-21)

The ultimate goal is to lay up treasures (reward, spiritual value) in heaven. All of these examples involved hypocrites having their treasure here while Jesus was saying for us to have our treasure in heaven, more specifically, in a spiritual realm. Again, this was not about salvation.

Remember, you can bring your spiritual value into the physical realm through prayer, whether it was like Hezekiah or to have a bigger house, etc. However, Jesus said to accumulate spiritual value through your righteousness (e.g., charity, prayer, and fasting) and not to bring it into the physical realm. Now we are ready to look at God's individual will.

Individual Will

> *Then the word of the Lord came unto me, saying, Before I formed thee in the belly I knew thee; and before thou camest forth out of the womb I sanctified thee, and I ordained thee a prophet unto the nations.* (Jeremiah 1:4-5)

> *For by grace are ye saved through faith; and that not of yourselves: it is the gift of God: Not of works, lest any man should boast. For we are his workmanship, created in Christ Jesus unto good works, which God hath before ordained that we should walk in them.* (Ephesians 2:8-10)

We saw how these two passages showed God could know every person and have a plan for each person that involves good works to generate spiritual value. This means, God has a specific plan for each person that is ideal. How God wants to bring about that plan for the individual is God's will for the person.

This individual will is unique to the individual (given at conception) and at birth has a finite maximum spiritual value (perfect) that is unique to the individual. The individual's ability to achieve God's perfect individual will is dependent on how much direction (grace) they take from God after they are born.

> *Then said Jesus unto them, When ye have lifteth up the Son of man, then shall ye know that I am he, and that I do nothing of myself, but as my Father hath taught me, I speak these things. And he that sent me is with me: the Father hath not left me alone; for I do always those things that please him.* (John 8:28-29)

Jesus achieved infinite spiritual value because He achieved the maximum by *only* doing everything in grace, and His life was unjustly cut short when He could have lived forever. We know that Jesus gave His infinite amount of spiritual value to God to facilitate our ability to attain salvation. When it comes to reward and us, we all know we aren't going to achieve the maximum because we don't do everything through grace. What happens when we do something apart from grace and it ends up as a negative value?

And we know that all things work together for good to them that love God, to them who are the called according to his purpose. (Romans 8:28)

When we make mistakes, it is possible to lose spiritual value and even create a situation that is unprofitable. However, the Bible said to those who love God and follow His purpose (through grace), it is possible to take unprofitable situations and make them profitable; however, the result will be less than the perfect result. Notice, making an unprofitable situation profitable will also be less than if the individual had taken complete direction from God (grace) in the first place. Making it profitable won't be more value than the original maximum opportunity.

For example, God may have set up a situation where I could end up with a +10 spiritual value; however, my actions made it a -10. If I love God and follow His leading me through grace, He can tell me what to do to make the -10 into a +1. At this point, God's will is something less than what was originally intended because of me, not because of God.

A Personal God

Which card player is more awesome? The one who wins according to the rules or the one who has marked the cards and deals from the bottom of the deck? Declaring the cheating person more excellent than the one who wins by playing by the rules tells a lot about the individual and their heart.

When it comes to omniscience, which God is more personal? The God who only knows what everyone is actually going to do before they do it or the God who knows all the possible choices everyone can make, sends out an influence to everyone in order to help everyone attain the maximum spiritual value, and then recalculates everything for everyone moments later when He finds out for sure what everyone has chosen? Notice, this constant recalculating and adjusting on God's behalf makes God more interested and invested in our lives than a God who can know everything in advance.

Remember, God is right in *what* He does, and He is right in *how* and *why* He does it (Right-Right). Notice, a God that takes shortcuts in *how* and *why* He achieves *what* He wants makes that God Right-Wrong.

Qualifying For More

No one is perfect, so we are not going to achieve the maximum profitability God predestinated (originally intended) for us. However, there is a way for the individual to achieve more than what God calculates along the way. There is a way for a person who has made choices resulting in a maximum of 40% of what God originally intended to now be able to attain 50%. If others hinder God's will for the life of the individual, and he handles it well, he qualifies for something more through justice at the expense of the people who hindered God's will for the individual.

For example, God may want you to move to another town; however, let's say you are going to need a place to stay, a job, and a sponsor. God would search on your behalf for someone to state that they want a person to do a job in line with your ability. God would confirm this with the person. (Let's call them "Dave.") Then God would look for someone in the same town who is stating their will they would put a person up in their house. (Let's call them "Ed.") God may even initiate asking Ed if he would board someone. Then, God would find "Fred," the sponsor who will mentor you and/or give you money. When all three (Dave,

Ed, and Fred) have stated their will, God can tell you His will for you: to move to that city.

Notice, when you ask God what you should do, and you don't get an answer, it may be due to God not knowing for sure because He is still determining what is possible because of *other* people. It would be unjust of God to state a specific plan to you before the plan is in place. Also, it would be unjust of God to state a specific plan to you if it is not the right time, you don't have enough spiritual value, or you didn't learn what you were supposed to learn from your current situation. Let's assume you have qualified for this new opportunity.

If all three people follow truth, then everything works out fine. However, at any point, Dave, Ed, or Fred could change their mind from what they told God. The result is you would wonder what happened to God's will for your life. For example, "Why did the job fall through if God told me to move there?" You could spend years of your life wondering if you heard from God, or if God was wrong, or if God isn't all-powerful, or if God is evil.

In reality, the plan falls apart because of Dave, Ed, or Fred, and if you handle it well, it actually qualifies you for something even greater! You didn't cause the problem, so justice says you deserve more. The greater result is coming at the expense of the disobedient individual(s).

Look throughout the Bible. God moved for people who handled bad situations well. It looks as if God needed the person to go through suffering in order to be able to move for them. Noah, Joseph, Job, Moses, the prophets, Daniel, Jesus, Paul, etc. God can implement His will in response to justice when people handle suffering well because God can implement His will through people who have enough spiritual value.

Summary

Matthew 6 taught that people will have more and less reward based on their actions and attitudes. This can be due to positive actions initiated (caused) by us: feeding the hungry, giving to the poor, visiting those in prison, etc., and not drawing attention to ourselves. This can also be due to negative actions not initiated (caused) by us: people persecuting us, illness, etc., and handling the suffering well. Justice equals both of these scenarios in a manner that results in reward (spiritual value) to the individual.

In fact, I believe on Judgment Day someone will have more reward than someone else. The one with less reward could ask why they didn't get as much. The answer would be that the person with more reward had more problems they didn't cause, and they handled them well. I believe the first words out of the questioner's mouth will be: "I wish I had had more problems."

Dr. Joel Swokowski's Commentary

All Things for Good

And we know that all things work together for good to them that love God, to them who are the called according to his purpose. (Romans 8:28)

A deeper look into this verse also shows another flaw in Calvinism. First, when Paul stated "all things," it meant God is able to work all things, not some things. Second, going back to Romans 8:18, we see that we must face the "sufferings of this present time." With that said, the "all things" worked together for good includes bad and/or wrong things. Isn't that the point of this verse? To encourage us that even when we make a mistake, God can fix it for good?

Point being, if God can turn our mistakes into good, why is God having to adjust at all? According to Calvinism, wouldn't He have seen the mistake before it happened? Wouldn't the mistake have been the plan all along? Wouldn't making up for our mistake really show a flaw in the Calvinists' belief in God's plan from the start?

> *As it is written, Jacob have I loved, but Esau have I hated.* (Romans 9:13)

Now we can see that God set Jacob up to be able to achieve more spiritual value than Esau; however, what each actually attains is dependent on how closely they follow God's will for each. God doesn't have a negative will for a person. God doesn't intend for a person to finish with negative spiritual value. Also, just because someone is able to attain more spiritual value than another person, it doesn't mean it is easier for the person to end up ahead (Luke 12:48).

We saw that spiritual value (reward) has a role in determining God's response to your will. Do people with more spiritual value continue to store it in spiritual places, or do they bring it into the physical world for an easier life now?

CHAPTER 9

Group Wills

SYNTHESIS SYSTEMS THINKING allows us to look at a group of parts as if it was a single part. For example, all of the children in a second-grade class can be spoken of as Mrs. Brown's second-grade class. Likewise, all of the second-grade classes can be spoken of as Springfield Elementary's second grade. Basically, synthesis systems thinking allows a group of people (members) to be seen as an individual (one body).

We saw God did this when He spoke to Jeremiah about Judah and Israel as women He married or when He spoke about loving Jacob and hating Esau. In fact, God sees marriage between two people as if it is one person.

Systems thinking treats a group as an individual, so it is easier to talk about the interaction between two groups as well as how to help a group of people. This means everything we covered in the last chapter also applies to a group of people, which we are calling God's group will. Now, we can look more closely at what was going on with each Bible passage covering God's will.

Moses

God approached Moses with regard to His group will concerning Israel: He would wipe them out and build a greater nation from Moses. Moses approached God with regard to Moses' individual will and said he personally didn't want it and that he had enough spiritual value to change God's Mind/will. However, that wasn't the end of the story. Here is what Moses said just before Israel occupied the promised land:

> *For I know thy rebellion, and thy stiff neck: behold, while I am yet alive with you this day, ye have been rebellious against the Lord; and how much more after my death? Gather unto me all the elders of your tribes, and your officers, that I may speak these words in their ears, and call heaven and earth to record against them. For I know that after my death ye will utterly corrupt yourselves, and turn aside from the way which I have commanded you; and evil will befall you in the latter days; because ye will do evil in the sight of the Lord, to provoke him to anger through the work of your hands.* (Deuteronomy 31:27-29)

Moses essentially stated God was right about the people, and it took Moses forty years to see it for himself. God's will was better than Moses' will.

Jeremiah

God approached Jeremiah with regard to His group will concerning Israel: even if Moses or Samuel were there to exert their individual will, they didn't have enough spiritual value to get Him to change His Mind/will.

Ezekiel

God approached Ezekiel with regard to His group will concerning Israel: even if Noah, Job, or Daniel were there, they would only be able to save themselves by their righteousness and not their wives or children with their individual will.

Hezekiah

God approached Hezekiah with regard to His individual will concerning Hezekiah: Hezekiah's illness would end in death. Hezekiah approached God with regard to Hezekiah's individual will: he didn't want to die. He had enough spiritual value to change God's Mind/will, which led to God not being able to achieve His group will with respect to Judah. Hezekiah's expression of his individual will made God's group will for Judah unprofitable.

Implication

A Super Bowl winning coach, who professes to be a Christian, said that God doesn't care who wins a football game. What do you think of that statement?

First of all, I always find this comment interesting from the standpoint that Jesus said in Matthew 10:30: *But the very hairs of your head are all numbered.* God cares about how many hairs you have on your head to the point He knows the number, but He doesn't care who wins a football game? It would seem hypocritical for a Christian who believes this to listen to or watch sports more than they listen to or watch the

"Hairs of Your Head" show. It also sounds like a God that is something less than personal.

More importantly, if God doesn't care who wins a football game, then God doesn't have a group will for football teams. Worse yet, God doesn't have an individual will for the football players, their wives, family, friends, and relatives when it comes to an activity that takes up a significant portion of their time. Is God unable to account for the complexity a football game brings to His creation? This belief sounds like they are saying God is rather small and limited in His ability.

Perhaps this coach was trying to say God wants both teams to win, so He doesn't care who wins because the result doesn't matter. However, the result makes a huge difference in the lives of everyone involved in the game, including their wives, kids, relatives, friends, fans, etc.

Is God choosing not to account for the result of the game in His will for the individual and the groups affected by the outcome of the game? This sounds like God has the ability, but He is taking this portion of His job less seriously and letting it affect all of His other plans.

What if God did have a will for both teams?
What if God wanted one team to win?
What if God wanted the other team to lose?

If your thought was: "Why would God have a will for a team to lose?," you have been affected by the culture and man's way of thinking. Why do people think it is always God's will for someone to win? Does winning in the short term always lead to more progress towards God? Does it always lead to more reliance on God? Does it always result in more reward in heaven? Does it always result in profitability? Which "profitability" do you think God is measuring? Can losing be a good thing?

Here is the mind-blowing question: What if God's will was for one team to lose, and they ended up actually winning the game?!?!

Take a second or two and reread that question, being conscious of the thoughts that make you uncomfortable.

If your thought was: "A team wouldn't be able to win if God wanted them to lose," you have been affected by the culture and man's thinking! You believe in the traditional view of an omnipotent God, and that He can force His will on us. This belief would result in Jesus' words to pray for God's will to be done on earth as it is in heaven proving to us He is not the Son of God!

The only non-contradictory conclusion is that God does care who wins the football game, but the result doesn't prove God's will.

What if God wanted a team to lose, and they won?

When you look at it this way, the worst thing that can happen is for God's will to be that a team loses, they end up winning, and the culture causes the team to think they are in God's will when they aren't! We will cover measures for being in God's will in Book 2.

Universal Will

It turns out, there is one more group will, and it was implied by our look at the dispensations. God has a will for His eternal plan relative to the human race as a whole. I call this God's universal will. Remember this?

The only way for Jesus' government and peace to grow in eternity is that value needs to be continually generated and this will be done by the interaction of beings in their uniqueness with the right and just Holy Spirit flowing through them via grace. This is God's plan!

We saw the Bible ended with believers occupying the new Jerusalem, which was called the tabernacle of God. This can be seen as the Church. We also saw the new Jerusalem was referred to as the Bride of Christ. This can be seen as the Marriage.

The meaning of life is church and marriage since this is how we will spend eternity (Ephesians 4 and 5). Notice, both of these are group wills. Both of these are supposed to generate spiritual value for everyone involved by the interaction of the individuals in their uniqueness with the right and just Holy Spirit flowing through them by grace.

Church and marriage were created by God, and they are God's way of generating spiritual value for everyone and this spiritual value ought to bring about God's universal will. This makes church and marriage both God's plan and God's will!

What are the two things the enemy attacks? Churches and marriages! This is how the enemy opposes God's plan and will. We will look at church and marriage in Book 2.

Summary

What is God's will?

God's will is made up of three general wills:
- Individual will
- Group will
- Universal will (Meaning of life)

God has a will in each area that can be attained if truth is followed. These three wills don't contradict each other.

God has a will for every individual and, at birth, has a finite maximum spiritual value ("perfect") that is unique to the individual. The individual's ability to achieve God's perfect individual will is dependent on how much direction (grace) they take from God. No one is perfect, so we are not going to achieve the maximum profitability God predestinated (originally intended) for us. However, there is a way for the individual to achieve more than what God calculates along the way based on how they handle injustices done against them by others.

God has a will for a group of people who exist at the same time. Whether it is a family, a church, a school, the Israelites in Egypt, or a football team, God has a plan for a group that can be attained if truth is followed and can be exceeded if the group handles adversity well that it didn't cause. Again, the spiritual value is attained through justice at the expense of others.

If the group will can be seen as a bigger system than the individual will, then the third will of God is the biggest system, and it is based on the smaller systems. God has a will for His eternal plan and the perfect version won't be attained because His will for individuals failed:

> *For this is good and acceptable in the sight of God our Saviour; Who will have all men to be saved, and to come unto the knowledge of the truth.* (I Timothy 2:3-4)

God's will for every individual is that they achieve salvation and come to the knowledge of truth. Notice, God's will for the individual and the group can grow. His original universal will is always running down.

He will never achieve His perfect universal will because people chose not to come unto the knowledge of the truth.

God's will happens through people who have enough spiritual value.

We have seen this means we affect whether and how much of the individual and group will occur. However, the Book of Revelation shows God's universal will is going to happen. This means, no one can stop God's universal will from happening. Rather, we are responsible for *when* God's universal will occurs and at what *cost*. It is not God's fault that the longer His universal will takes to happen, the more people will spend eternity in the lake of fire.

God is right and just in *what* He wants, *why* He wants it, and *how* He goes about bringing it to pass (Right-Right). The meaning of life is church and marriage, and God gave us church and marriage to bring about His universal will. Before we look at that in more detail, we need to understand another dimension of God's wills.

It looks as if the latest image is coming into even more focus! We can now see the image of God working through a human (on the left side of the wall) and a group of people (on the right side of the wall) resulting in a Bride made up of a group of people that is married to Jesus.

Dr. Joel Swokowski's Commentary

Universal Will

We already saw that I Timothy 2:3-4 teaches that God would have all men to be saved.

Here are a few more verses that emphasize that same point:

> *For God so loved the world, that he gave his only begotten Son, that whosoever believeth in him should not perish, but have everlasting life.* (John 3:16)

> *The Lord is not slack concerning his promise, as some men count slackness; but is longsuffering to us-ward, not willing that any should perish, but that all should come to repentance.* (2 Peter 3:9)

> *Have I any pleasure at all that the wicked should die? saith the Lord God: and not that he should return from his ways, and live?* (Ezekiel 18:23)

> *O Jerusalem, Jerusalem, thou that killest the prophets, and stonest them which are sent unto thee, how often would I have gathered thy children together, even as a hen gathereth her chickens under her wings, and ye would not!* (Matthew 23:37)

How many other verses would a person need to see that God's universal will is that all people would be saved? Do you see how quickly this point shows the flaw in Calvinism?

If God wants all to be saved, what are the two most crucial doctrines to understand?

God's Nature and salvation, and now we're back to the importance of the information presented in *Modeling God*. Understanding God's Nature and salvation are foundational to everything we are learning in this book.

Church and Marriage

Let's not look too quickly past the hugely important point made in this chapter: church and marriage are the meaning of life. Simply put, we were all made to love one person and be loved by one person (individual will), and to love a community and be loved by a community (group will). God's universal will accounts for both of these in that the *Church* will be *married* to Christ for eternity!

> *Therefore as the church is subject unto Christ, so let the wives be to their own husbands in every thing. Husbands, love your wives, even as Christ also loved the church, and gave himself for it; That he might sanctify and cleanse it with the washing of water by the word, That he might present it to himself a glorious church, not having spot, or wrinkle, or any such thing; but that it should be holy and without blemish.* (Ephesians 5:24-27)

Church and marriage are the vessels through which God's will happens. Are the churches and marriages you see bringing about God's will, or hindering it?

CHAPTER 10

Three Levels in God's Will

And be not conformed to this world: but be ye transformed by the renewing of your mind, that ye may prove what is that good, and acceptable, and perfect, will of God. (Romans 12:2)

THERE IS DEBATE on whether Paul was saying God's will is always good, acceptable, and perfect or God's perfect will is also good and acceptable. The latter option would also lend itself to the belief a person in God's will is at one of three levels: perfect or acceptable or good.

Whether Paul was stating that last option or not, I'm going to show you that it is, in fact, possible to be in God's will at one of three levels: good, pleasing (acceptable), or perfect.

After all, we have seen that God had a will for Moses that was perfect, yet Moses' prayer resulted in God supporting Moses' plan. The Israelites did make it to the promised land by being in God's will, but it was much less profitable than how God had originally intended. How would you describe these two wills?

Furthermore, the above verse from the Book of Romans shows that a person who is transformed is able to determine what is the will of God! No wonder understanding God's will (the biggest system) leads to being able to explain anything! Ultimately, this edition is trying to help you be transformed by the renewing of your mind, which is what the Holy Spirit wants for you. This means the individual experiences tension (stress) as an effect of wanting to do their own will instead of God's will. Let's look at these levels.

Perfect Will

God has an approach to bring about His will that is perfect. We saw perfect means maximum profitability. In fact, we saw an example of this in the last chapter:

> *For this is good and acceptable in the sight of God our Saviour; Who will have all men to be saved, and to come unto the knowledge of the truth. (I Timothy 2:3-4)*

As it relates to God's universal will, His perfect level would have resulted in everyone being saved and coming to the knowledge of the truth. As it relates to God's individual will, we already saw God has a perfect will for each individual before they are born and no one except Jesus ever achieves it. If a person believes the only level of God's will is perfect, then they believe that no one other than Jesus is in God's will.

What does it look like to be in the perfect will of God in the moment?

> *Now the Lord had said unto Abram, Get thee out of thy country, and from thy kindred, and from thy father's house, unto a land that I will shew thee: And I will make of thee a great nation, and*

> *I will bless thee, and make thy name great; and thou shalt be a blessing: And I will bless them that bless thee, and curse him that curseth thee: and in thee shall all families of the earth be blessed. So Abram departed, as the Lord had spoken unto him; and Lot went with him: and Abram was seventy and five years old when he departed out of Haran.* (Genesis 12:1-4)

God told Abram what to do (grace), and Abram immediately did it (faith). In fact, this faith was credited to Abram as Righteousness for salvation because of grace: the divine influence upon his heart, and its reflection in his life.

> *By faith Abraham, when he was called to go out into a place which he should after receive for an inheritance, obeyed; and he went out, not knowing whither he went.* (Hebrews 11:8)

> *Cometh this blessedness then upon the circumcision only, or upon the uncircumcision also? for we say that faith was reckoned to Abraham for righteousness.* (Romans 4:9)

As it relates to God's group will, the same could be said for a group. God told Moses to tell the people to prepare food a specific way and to ask the Egyptians for valuables before He freed them from Egypt. Then, when Jehovah struck Egypt with the final curse:

> *And the people took their dough before it was leavened, their kneadingtroughs being bound up in their clothes upon their shoulders. And the children of Israel did according to the word of Moses; and they borrowed of the Egyptians jewels of silver, and jewels of gold, and raiment: And the Lord gave the people favour in the sight of the Egyptians, so that they lent unto them*

such things as they required. And they spoiled the Egyptians. (Exodus 12:34-36)

Acceptable Will

The word acceptable (G2101 "euarestos") in this verse also translates to "fully agreeable and well pleasing." I like to refer to this as God's pleasing will because He is still excited about this, whereas, some people see acceptable as God allowing it and, because it's less than perfect, people feel bad about it. This will is pleasing to God! Notice, God's perfect will is also pleasing to God.

When it comes to God's individual will in the moment, this means, God tells the individual what to do through grace and the person doesn't immediately do this. Basically, the person feels the tension of choosing to do it and it can feel like it is too much for them to bear. They need some help getting over the tension, whether it is more understanding or experience, they need more faith. Gideon is a great example of God's pleasing will.

And Gideon said unto God, If thou wilt save Israel by mine hand, as thou hast said, Behold, I will put a fleece of wool in the floor; and if the dew be on the fleece only, and it be dry upon all the earth beside, then shall I know that thou wilt save Israel by mine hand, as thou hast said. And it was so: for he rose up early on the morrow, and thrust the fleece together, and wringed the dew out of the fleece, a bowl full of water. And Gideon said unto God, Let not thine anger be hot against me, and I will speak but this once: let me prove, I pray thee, but this once with the fleece; let it now be dry only upon the fleece, and upon all the ground let there be dew. And God did so that night: for it was

> *dry upon the fleece only, and there was dew on all the ground.*
> (Judges 6:36-40)

Gideon didn't immediately follow God's influence. Instead, he asked for two signs. Was God pleased with this?

> *And what shall I more say? for the time would fail me to tell of Gedeon, and of Barak, and of Samson, and of Jephthae; of David also, and Samuel, and of the prophets: Who through faith subdued kingdoms, wrought righteousness, obtained promises, stopped the mouths of lions. Quenched the violence of fire, escaped the edge of the sword, out of weakness were made strong, waxed valiant in fight, turned to flight the armies of the aliens.*
> (Hebrews 11:32-34)

Hebrews 11 has been called "The Faith Hall of Fame" and listed examples of people in the Old Testament who demonstrated great faith. Gideon being used as an example to us showed he was in God's will; however, what he did wasn't perfect, so I see this as pleasing.

It looks as if people in God's pleasing will need someone outside of themselves to help them overcome the tension. They need a leader. It appears the profitability from doing God's will is shared among the person and their leader. In the Gideon example, God was Gideon's leader. We can have a share in the profitability of others if we can encourage others to do God's will.

Again, the same could be said for God's group will. Goliath taunts Israel for forty days. David ends up serving as leader to Israel by slaying Goliath (I Samuel 17).

When it comes to God's universal will, when Eden failed, God tried to begin the process with Noah, and then with Abraham. As for God's perfect will (that all are saved), it doesn't happen, but God is pleased with those who do get saved.

Good Will

We have seen good means to create something of value. For example, we've seen that even when we initially oppose God and do our own unprofitable will, if we love God and repair through grace, something good can still result. This will happen if we love God and repair the situation by following God's direction through grace, yet we will never exceed His perfect plan and will. Jonah is a great example of this.

> *Now the word of the Lord came unto Jonah the son of Amittai, saying, Arise, go to Nineveh, that great city, and cry against it; for their wickedness is come up before me. But Jonah rose up to flee unto Tarshish from the presence of the Lord, and went down to Joppa; and he found a ship going to Tarshish: so he paid the fare thereof, and went down into it, to go with them unto Tarshish from the presence of the Lord. (Jonah 1:1-3)*

If Jonah was in God's perfect will, he would have immediately gone to Nineveh. Instead, he went the opposite direction.

> *But the Lord sent out a great wind into the sea, and there was a mighty tempest in the sea, so that the ship was like to be broken. Then the mariners were afraid, and cried every man unto his god, and cast forth the wares that were in the ship into the sea, to lighten it of them. But Jonah was gone down into the sides of the ship; and he lay, and was fast asleep. (Jonah 1:4-5)*

God increased the tension to try and get Jonah to do God's pleasing will; however, Jonah continued to oppose it. The men of the ship cast lots and God showed them this was Jonah's fault. When they woke Jonah, and he confirmed it, they responded:

> *Then said they unto him, What shall we do unto thee, that the sea may be calm unto us? for the sea wrought, and was tempestuous. And he said unto them, Take me up, and cast me forth into the sea; so shall the sea be calm unto you: for I know that for my sake this great tempest is upon you.* (Jonah 1:11-12)

Jonah was essentially saying he'd rather die than do God's will! The men didn't want to kill him, but when there was no other option, they threw Jonah overboard and...

> *Then the men feared the Lord exceedingly, and offered a sacrifice unto the Lord, and made vows. Now the Lord had prepared a great fish to swallow up Jonah. And Jonah was in the belly of the fish three days and three nights.* (Jonah 1:16-17)

It appeared it took Jonah spending three days and nights in the belly of the fish with seaweed wrapped around his head (Jonah 2:5) for him to agree to God's will and pray to Jehovah so that the fish vomited him on the shore towards Nineveh!

Notice, Jonah could have decided to remain outside of God's will; however, God increased the tension to such a degree that Jonah chose to do God's will. Jesus referred to Jonah as a prophet, but would you call this situation perfect? Would you say God was pleased by Jonah's actions? All you can really say is that Jonah did the bare minimum to do God's will, and he can't be compared to Abraham, nor Gideon. Do you think Jonah should be in The Faith Hall of Fame?

When it comes to a group will, the Israelites end up walking the desert for forty years before they enter the promised land. They were still in God's good will, but He was not pleased, and it was far from perfect. However, we can now see that God's perfect will and His pleasing will are also good.

When it comes to God's universal will, we saw He will create a new heaven and new earth with a new Jerusalem; however, many people will end up in the lake of fire after God brings the ultimate tension with the Tribulation. God's perfect will is that all are saved. While God is pleased with people getting saved, what would His good will look like? Wouldn't it be people making a deathbed confession or people having to hit rock bottom (like Jonah) before they renounce their plan for their life and allow God to use them to bring about His will? What do you think of the evangelist who is able to help people get saved *before* these people's lives hit rock bottom?

Out of God's Will

People who do their own will instead of God's will are unprofitable. They end up destroying, which we learned is the definition of evil. We can see this as someone who opposes the maximum tension of God's good will. Pharaoh was a great example of this. God brought the maximum tension with the last plague, and Pharaoh initially responded to God's good will by setting the Israelites free.

> *And it was told the king of Egypt that the people fled: and the heart of Pharaoh and of his servants was turned against the people, and they said, Why have we done this, that we have let Israel go from serving us? And he made ready his chariot, and took his people with him: And he took six hundred chosen*

chariots, and all the chariots of Egypt, and captains over every one of them. (Exodus 14:5-7)

Pharaoh chose to do his own will and it ended up catastrophically unprofitable for him when God closed up the Red Sea over Pharaoh's army.

When it comes to a group will, God had a will for the Israelites coming out of Egypt and it was to go into the promised land after sending in twelve spies. Two of the spies (Caleb and Joshua) were in the perfect will of God. It was said the other ten gave an evil report (Numbers 13:32) because they convinced the people to do their own will.

When it comes to God's universal will, some people are going to ignore God's good will and do their own will to the point it sears their conscience and all that is left for them is the lake of fire.

Summary

The Ultimate Answer: The Three Plans.

1. God has a specific and perfect plan for each individual who has ever existed.
2. God has a specific and perfect plan for each group of people (a plan for a group of people who exist at the same time).
3. God has an ultimate plan for the human race that is going to happen (the meaning of life); however, it's up to us whether we're part of it or not.

God's will is seen relative to these three general plans: individual, group, and universal.

God has a plan in each area that can be attained if truth is followed. These three plans don't contradict each other. The individual and group plans facilitate the universal plan.

Notice, God's plan for the individual and for the group can grow but can never exceed His perfect and original plan: that all are saved on this earth, and the Jerusalem He did create would have been *the* Jerusalem.

When it comes to the individual relative to their spiritual growth, the evil will is the lake of fire. The good will is salvation. The pleasing will is sanctification. The perfect will is discipleship, which were Jesus' last words on earth.

When it comes to a group, the evil will is being cursed. The good will is churches following truth but not working together. The pleasing will is churches working together and operating according to the gifts Jesus provided in Ephesians 4 (apostle, prophet, evangelist, pastor, and teacher). The perfect will is when we are in the new Jerusalem acting as one body.

When it comes to the universal will, the perfect will would have been for Adam and Eve to turn the earth into Eden by having descendants who completely followed God. Instead, God will get the same result in eternity, but not before a lot of unprofitability occurs.

God's perfect plan is running down due to the individual and groups not following truth, which results in the universal plan never including all being saved, and the longer it takes for individuals and groups to have enough value to bring about God's universal will, the more people will end up in the lake of fire. This is not God's fault.

As we continue to look at the wall, while we can't see the image at the top of the wall, it looks as if it is God. We began our journey away from the wall beginning in the middle of the wall at the bottom. The image at the bottom and middle of the wall is a person teaching another person to fish. However, above this image and below God, we see four levels. The top level is gold and includes God's perfect will. The level below it is silver and includes God's pleasing will. The next level is brass and includes God's good will. The lowest level is iron and includes being out of God's will. There look to be other images in each of these levels, so let's continue walking away from the wall until these come into focus!

Dr. Joel Swokowski's Commentary

God's Wills as a Tree

My friend and comrade, Jonathan Fries, taught me the following analogy as yet another way to break down the concept of God's good, pleasing, and perfect wills.

Think of God's good, pleasing, and perfect will like a tree with fruit. A good tree just means that it produces some fruit, even just one apple. It produced fruit, it is good. Even that one apple can provide sustenance. Perhaps there is an image of this tree on the brass level.

A pleasing tree would produce more fruit than expected. Perhaps there is an image of this tree on the silver level. There is an expectation for this tree to produce a certain amount of fruit, and a pleasing tree would exceed that expectation. The farmer (God) would be happy. Let's take a moment to understand happiness.

Happiness is when your brain releases pleasure chemicals in response to your context, and there is an equation for happiness: Happiness occurs when you experience an effect (result) that exceeds your expectations.

For example, if you were expecting $40 and I gave you $20, you would be upset. If you weren't expecting anything and I gave you $20, you would be happy. Notice, in both cases, I gave you $20. The only difference is what you were expecting.

God does have certain expectations of us, especially as leaders. When we exceed those expectations, God experiences happiness! I wrote in my commentary for *Modeling God* "Chapter 11. Love" that anger occurs when our expectations aren't met. Every time God was angry in the Bible, it was when God's expectations weren't met. Remember, God's expectations would always be right and just.

Now, people often confuse pleasing and perfect. We need to remember that perfect doesn't mean without flaw. Perfect means maximum profit. The perfect tree is the tree that produces the amount where there is no way for that tree to produce more fruit. Perhaps there is an image of this tree on the gold level. The Bible also refers to this as a 100-fold return (Mark 10:30). This is the life of Jesus, the perfect spotless Lamb, in every single moment.

So we can be those types of trees in every single moment of our lives:
- Is this a good moment? Did you produce any fruit?
- Is the fruit that you produced exceeding God's expectation of what He wanted you to do at that moment?
- Is the fruit produced the maximum amount possible at that moment?

Are you being like Jonah, Gideon, or Abraham?

Also, as I Timothy 2:3-4 showed, God's perfect will (that all are saved) is also pleasing and good. Likewise, Lenhart stated, God's pleasing will is also good.

Let's take this one step further. You can have the three scenarios presented above, but there's one missing. We talked about God's good, pleasing, and perfect will…but there is the option of just not being in God's will. That looks like any story where God's will doesn't happen.

As it relates to the tree analogy, this would be similar to Jesus seeing the fig tree with no fruit and cursing that tree (Mark 11:12-25). It's not necessarily wrong for a tree to not bear fruit (e.g., think of the winter season). However, the fig tree Jesus cursed was supposed to be bearing fruit and looked like it was bearing fruit, but it had no fruit. Jesus was expecting fruit and cursed it when His expectations weren't met. This is the option of not being in God's will. Perhaps there is an image of this tree on the iron level.

CHAPTER 11

Council Meetings

WE HAVE COVERED three types of wills God can have and three measures for the level of God's will that is occurring. We will close out Book 1 with characteristics that will help you recognize God's will vs. your will, so that you will be able to accomplish the goal of Book 2: recognizing God's will. Before we cover the first characteristic, I want to cover the difference between a leader and a wiseman (or wisewoman).

Leader vs. Boss

What is your definition of leader?

A definition ought to be a cause; however, the overwhelming majority of definitions, when they are provided, are effects. For example, a very popular definition for leadership is influence, nothing more and nothing less.

I love to ask these people, "If a person slipping on a banana peel influences you to avoid that path, would you tell others he is your leader?"

Notice, all explanations, excuses, qualifiers, etc., in an attempt to rationalize this abusive definition by being more specific, actually means that leadership is influence *and* something more (explanation).

Another example occurs when I ask people who believe this definition (leadership = influence) and are presenting at a Leadership Conference if they consider Hitler to be a leader. When they say, "Yes," I tell them, "Good luck with your Hitler training."

They quickly respond, "Hitler was a *bad* leader" and I simply point to their banner and say, "Then why doesn't it say *Good* Leadership Conference?"

Believe it or not, the worst example is when people say "servant leadership." A healthy brain works from cause to effect. When people can't give a definition as a cause leading to the effect they want, they take the effect and make it part of the definition, and this ought to be the first sign to you that the person is likely a dictator teaching "leadership."

What is servant leadership? (Not Hitler)

What does non-servant leadership look like? (Hitler)

This is the worst example of abuse because not only are we open to accepting this damage to our brain, we are energized to spread it to others!

Leader: a person who facilitates the purpose and progress **of others**.

The limitation in this conjunctive is **bolded**. Notice, the effect of this definition is that the leader is a servant. What is non-servant leadership?

Boss: a person who facilitates their own purpose and progress **at the expense of others**.

Here are just three applications of these definitions:

- Leaders bear pain, they don't inflict it. Bosses inflict pain, they don't bear it.
- Leaders like to be led. Bosses don't like to be bossed.
- Everyone can be a leader. Only one person can be the boss.

Notice, God is a leader; He is sufficient in Himself, so His focus is facilitating your purpose and progress. When people know their purpose and are making progress in it, they don't have any issues. Everyone who needs help needs it because they either don't know their purpose or they aren't making progress.

When you have these definitions, it becomes immediately obvious whether a pastor, husband, king, etc. is a boss or a leader. In the Bible, every king that was declared good was a leader and not a boss. Remember the wall analogy from the preface of *Modeling God*? Leadership is being represented by the image of a man helping another man learn how to fish.

Remember, God being sufficient is different from God's will. In order for God to bring about His plan, which involves humans, He is going to need humans. In Book 2 we will look at this more closely.

Wise Man

In the Bible, whenever the leader had a problem they couldn't solve, they looked for a "wise man." This was a person who spent their time in deep thinking and was able to solve problems. They were often experts in specific subject areas or had specific gifts. The more successful the leader, the more wise men they had. For example, in Daniel 2, Nebuchadnezzar has a dream that troubled him and caused him not to be able to sleep. When he finds that all his wise men couldn't tell him what the dream was and its interpretation, verses 12 and 13 said:

> For this cause the king was angry and very furious, and commanded to destroy all the wise men of Babylon. And the decree went forth that the wise men should be slain; and they sought Daniel and his fellows to be slain.

Daniel ends up resolving the king's problem and gets promoted. Notice, the king not only doesn't have a problem not being the wisest person in his kingdom, he knows being the wisest man is not his job. He knows a leader can't spend the time required to be the wisest person *and* still be an effective leader.

This means the leader and the wisest person ought to be two different people. The Bible is filled with examples: Pharaoh and Joseph, Nebuchadnezzar/Darius and Daniel, David and Samuel/Nathan/Gad, Jesus and Elijah/Moses, and even Saul and Samuel, until Saul decides to do both jobs, which is the same mistake Solomon, Rehoboam, and Satan made. Even in eternity, Jesus is the leader and God the Father is the wise man.

Councils

There is an eye-opening passage in the Bible where we get to see a step-by-step process for how God determines His will, and we needed to understand the roles of leader and wise man before we covered it. Here is I Kings 22:

> *And he said, Hear thou therefore the word of the Lord: I saw the Lord sitting on his throne, and all the host of heaven standing by him on his right hand and on his left.* (I Kings 22:19)

What's going on here? The host of heaven are the spiritual beings (angels?). The word *host* comes from a word often translated as army (H6635 "saba"); however, that word is also used to mean service and to serve at the sacred tent.

> *And the Lord said, Who shall persuade Ahab, that he may go up and fall at Ramothgilead?* (I Kings 22:20a)

God presented the agenda item: "There's enough justice against Ahab that it's time for Ahab to die. The issue is that Ahab needs to be convinced to go to Ramothgilead. How should we do this?" This is a council meeting. A council is an assembly of persons summoned or convened for consultation, deliberation, or advice.

> *And one said on this manner, and another said on that manner.* (I Kings 22:20b)

Do you see what is going on here? God is the leader who has called this meeting and will make the final decision. The host are acting as wise men and offering their answers. It's like a brainstorming session.

Apparently, none of the answers are good enough, but no one is getting punished. Why doesn't God come up with the answer? Besides the fact that God would be acting as leader *and* wise man, it looks as if God is doing this to give the host the opportunity to grow. Everyone needs to grow up to God!

> *And there came forth a spirit, and stood before the Lord, and said, I will persuade him.* (I Kings 22:21)

The host is made up of spiritual beings, and one of them said he would persuade Ahab.

> *And the Lord said unto him, Wherewith?* (I Kings 22:22a)

Again, God stayed in the role of leader and asked the spirit acting as wise man what he would do.

> *And he said, I will go forth, and I will be a lying spirit in the mouth of all his prophets.* (I Kings 22:22b)

What a strange and unorthodox plan. God would never agree to this, would He?

> *And he said, Thou shalt persuade him, and prevail also: go forth, and do so.* (I Kings 22:22c)

Notice, God didn't come up with this idea and He wasn't going to carry it out. God acted as leader and picked an idea that was not His! Wait, I thought the measure of maturity was that you come up with ideas to solve problems on your own? Well, not according to God!

> *Now therefore, behold, the Lord hath put a lying spirit in the mouth of all these thy prophets, and the Lord hath spoken evil concerning thee.* (I Kings 22:23)

Micaiah said God put the lying spirit in the mouth of the prophets. Why did he say that? It wasn't His idea, yet, it was His decision and His responsibility as leader!

If we imagine a problem is like a small gold "x" painted in an enormous, intricate painting. The answer is the position and perspective needed to see the gold "x." It can take days to determine the right position and perspective to view the painting. However, as with all problems, once you have determined the answer, not only is the answer hard to forget, it causes you to wonder why you hadn't seen it before. In fact, you can't not see it!

The reason the solution is difficult to see is it takes humility and contrastive thinking. However, a leader must show confidence and determine a way for everyone to gain, otherwise he creates a distraction due to the follower's doubt. It is exhausting to try to be the expert in both leading and determining the solution.

Instead, a good leader identifies as many wise men and women as possible, cultivates strong relationships with them so that they want to share with the leader, and allows them the freedom to look at the issue from every position and perspective until they find potential solutions. Then, the king allows them to share, knowing that not only did he save time determining the solution, but his job is to make the decision, not necessarily to come up with the decision.

Council Meeting Basics:

- A council meeting has a leader: a person responsible for any decision that could be made.
- The leader puts out an objective/agenda to the council members.
- The council gives their thoughts, all their thoughts, whether they are good or bad, so the leader has a wide range to choose from.

- Once there are enough stated thoughts/ideas (it could take more than one meeting), then the leader endorses an idea and makes sure it happens.

And God said, Let us make man in our image, after our likeness: and let them have dominion over the fish of the sea, and over the fowl of the air, and over the cattle, and over all the earth, and over every creeping thing that creepeth upon the earth. (Genesis 1:26)

And after six days Jesus taketh Peter, James, and John his brother, and bringeth them up into an high mountain apart, And was transfigured before them: and his face did shine as the sun, and his raiment was white as the light. And, behold, there appeared unto them Moses and Elias talking with him. (Matthew 17:1-3)

Now we can see that creation (Let *us* make man in *our* image... Genesis 1:26), the Transfiguration of Christ, as well as many of the teaching interactions Jesus had with His disciples, were all council meetings.

Husbanding

I am the true vine, and my Father is the husbandman. Every branch in me that beareth not fruit, he taketh away: and every branch that beareth fruit, he purgeth it, that it may bring forth more fruit. (John 15:1-2)

Jesus called God His husbandman. What is a husbandman? Do you know the definition of husbandry?

Husbandman comes from two Greek root words (G2041 "ergon" - work, G1093 "ge" - land) that, when combined, means worker of the land. Husbandry is the care, cultivation, and breeding of crops and animals. A husband works the land in order to make it profitable. Notice, you don't have to be married in order to be a husbandman.

In light of the definition of leadership, a husband is someone focused on facilitating the purpose and progress of a specific person for an extended period of time. They are focused on growing the person into everything they are capable of.

God cannot grow. The only way God can interact with others while being completely Himself is that others have to grow up towards God. This is why God is completely focused on growing everyone, even Jesus. I see council meetings as another opportunity for God to husband others by putting them in the role of wise man.

On an archery target, the center is called the bull's eye. Around the bull's eye is a small ring called the inner ring. The way I look at God's will is that the bull's eye is God's perfect will. The inner ring is God's pleasing will. Any other shot hitting the target outside of the inner ring is God's good will.

When we look at the story from Micaiah, we can see the spirits taking their shot at the target and God not agreeing to their idea until it at least hits somewhere inside the inner ring. Basically, God as leader is going to make sure that the decision He goes with is not wrong or unjust, and that makes God the safest person to share with. When we see sharing with God as an opportunity to grow, why wouldn't we want to share with God? Remember, sharing is completely different than stating your will through prayer.

We saw in *Modeling God* that I Samuel 8 gave us a view of prayer from God's perspective. Basically, you share your request and then God shares His insight. However, the choice for what and how you pray is up to you, and you make the final decision, after hearing God's feedback. When we consider council meetings, we can now see that prayer puts you in the position of leader and God in the position of wise man, if you take time to listen to Him during the prayer. No wonder prayer has an impact on God's will, it is the same process as God uses to determine His will!

Summary

Can you see how council meetings are a characteristic of God's will and not God's plan? Council meetings are literally *how* God makes the decisions needed to accomplish His plan!

Remember, God's plan is to hang out with us for eternity. God isn't focused on getting the perfect answer. He is interested in interacting with us in a manner that grows us closer to His level. Anyone who focuses on the *what* and making the effects perfect is hindering God's plan and proving they don't understand God's will.

In fact, when I speak to groups of people, I say there is one type of person who thinks they can only make decisions about solutions they themselves came up with. Do you know what we call these people? Teenagers.

We are all the leaders of our own lives, so true maturity is shown by how many wise people we surround ourselves with because our success is determined by the options we choose from.

When you have an idea about how you want to go about achieving your plan, do you get council before making progress in that idea? What would be the reason you wouldn't?

God ought to be on your council via grace; however, why would a person refuse to share their plan and then rationalize it by saying the only person they need to talk to is God? After all, they are the leader of their life and what people say shouldn't stop them from doing what they want to do, unless they know their plan is wrong and/or unjust, then they wouldn't want to share a plan that doesn't even hit the target. That is proof it is man's will and not God's will.

How do you accomplish a project? Alone or with a team?

Who is on your council?

Dr. Joel Swokowski's Commentary

Lenhart presented three applications of the definition of a leader:

- Leaders bear pain, they don't inflict it. Bosses inflict pain, they don't bear it.
- Leaders like to be led. Bosses don't like to be bossed.
- Everyone can be a leader. Only one person can be the boss.

These are the effects of having non-contradictory definitions for leader and boss. They can be seen as shortcuts to determine if you are dealing with a leader or a boss. It will also help you determine if a group you are dealing with is made up of leaders or bosses.

King Arthur

When you think of the Arthurian legend, it's likely you quickly envision King Arthur and his Knights of the Round Table. This is a perfect depiction of a leader and a council group. King Arthur is often espousing, to his knights and to others, how everyone at that round table is equal. They are equal, while they are sitting at that table. They all have equal say over the ideas and strategies that are discussed during the session. Yet, once the meeting is concluded, when it's time to decide what plan will be executed, it is King Arthur, and King Arthur alone, who's responsible. The decision belongs to King Arthur. That burden lies with him, not with his knights. However, the wisdom and morality depicted by the Arthurian legend would not be what it is without the knights he surrounded himself with.

Creating a Council Group

Here are four things to do to set up a council group for yourself:

1. Determine the purpose of the group.
    a. General Council: getting help with your life in a broad sense. The agenda items would be limited by the areas *you* are dealing with.
    b. Specific Council: getting help with a specific project. The agenda items would be limited by the project itself.
2. Invite the people (two or more) you want on your council to this group with the following stipulations:
    a. Purpose of the council: Let them know the purpose of your council and what role you want them to play.
    b. Agreement Check: Ask them if they're in agreement with being on your council.

Keep in mind, this is your council. The people on your council should be people you trust to hear from God and speak grace into your life.

3. Once agreement is established, start bringing your agenda items to the council group. You are only limiting yourself when you don't bring something to the council group. If you think it's too small of a problem, then bring it anyway. Your council has agreed to help and should be excited to help you with such simple issues.
4. Determine the council group setting. These days, it has become quite simple to do this with all the virtual options. For instance, I've found setting up a group thread on some messaging app works excellent. You can even name the thread according to the council (Joel's Council, etc.).

It works, I've experienced it myself. When my son was around six years old, I decided to fight for more custody and placement rights. My goal was to reach a 50/50 split with his mother. As these things go, we had our disagreements and needed outside resources to resolve this conflict between us. We each pursued our own course of legal action within the court system.

My son's mother hired a lawyer, which served as her own version of a council group. Her lawyer was the #1 wise man, along with the team of other lawyers and paralegals at their law firm. A great picture of what a council group can look like and how it can be a benefit. That's the entire reason people hire lawyers; they are experts in their field, giving counsel on what they deem the best decision for you to make. The decision is still yours.

I chose to represent myself. This decision was partly due to my inability to afford a lawyer, but more so due to my faith in the council group

I could draw from in my community. I especially drew on the expertise and emotional and spiritual support of three people (two men and one woman) in my community. I had more than that pouring into my life and into this situation through prayer, but the people I was drawing on intentionally and proactively were what gave me the strength and courage to make it through this situation.

This journey brought me from the beginning of the mediation process facilitated by the county court system all the way through standing in front of a judge hearing his final judgment over the custody and placement of my son.

As you can imagine, the strain of a conflict like this can be trying... and it was. I had never been through anything like this before, and representing myself was nerve-wracking. However, the strength and wisdom from God through my council group gave me the energy and insight I needed to venture forward in what became a two-year process.

I was given the strategies I needed before each step of the process while still being encouraged to make the best decision for myself. I wasn't told what to do. I was given vision. I wasn't being controlled. I was empowered.

This path eventually led to my son being with me 50/50. It didn't happen the way or in the timing I thought it would, but it did result in three benefits I did not see coming:

1. I can look back on this experience and hold my head high. Not only did I handle a tense situation well and without causing any damage, but I also felt (and still feel) good about having fought for my son and for what was right and fair.

2. I am now able to and have been given opportunities to help other people going through similar situations. Walking this road for myself gave me an emotional outlook that is hard to replicate without having the experience. This has allowed me to be empathetic to others and offer the wise counsel I was given while going through this process.
3. I became closer than ever to the people on my council. There's nothing like going through a project together, even more so, walking through a conflict as a team. It's quite a bonding experience. I imagine that's one of the reasons God operates this way, His council meetings bring Him closer to those He's working with (and working through)!

CHAPTER 12

Dissolve

DR. RUSSELL ACKOFF liked to say there are four ways to approach a problem:

- **Absolve:** Ignore it, maybe it will go away. (If it's significant, it won't.)
- **Resolve:** Treat the symptoms/effects.
- **Solve:** Treat the tangible causes.

As we saw in the preface, problem solving is an analytical approach and works perfectly on inanimate objects (e.g., washing machine)—we take the problem apart, identify the cause, fix it, and put it back together. This makes sense because our brains naturally work this way by breaking things down into categories.

The problem is, whenever the solve approach is used on people, three more problems are created, which is known as the law of unintended consequences, and the overall stress of the solution is worse than from the original problem.

> *There is a way which/that seemeth right unto a man; but the end thereof are the ways of death.* (Proverbs 14:12 & Proverbs 16:25)

Solving people's problems seems right, but it actually results in making everything worse and, in the long term, unprofitable. This is why the world is getting worse. Think about all the laws that require us to pass more laws. What's the right approach?

- **Dissolve:** Treat the intangible causes.

Modeling God ended by contrasting people who saw the physical as the driver with those enlightened people who saw the spiritual as the driver. Now, we see it is easy to see who is on their way to being transformed: Is their immediate focus the physical cause or the intangible spiritual cause?

Again, as we saw in the preface, the synthesis approach is the way to completely understand anything, especially problems. Remember, the first step of the synthesis systems approach is to look away from the system we are dealing with and towards the bigger system, which is the opposite of how our brains naturally work. Likewise, the best approach to problems when it comes to people is to look away from the problem. This approach results in the problem never being directly addressed because it dissolves!

The reason we are addressing this as it relates to God's will is that, in the Bible, God never solved a problem when it came to people; God only dissolved problems. Furthermore, Jesus never saw the cause of any issue as tangible; He always saw the cause as intangible/spiritual. Jesus is our example of enlightenment and transformation!

That last point is one that has increased my faith in the Bible. I'm not aware of any other literary work where a character never sees the cause of a problem as tangible. Suppose the Bible was written by man, and man had created a character (Jesus) who always saw problems from a spiritual cause. Wouldn't there be at least one more character in all of literature who sees the spiritual as the cause of every problem people have?

This may be completely foreign to you, and you are most likely asking for examples. Let's look at a couple!

Solomon

> *7 In that night did God appear unto Solomon, and said unto him, Ask what I shall give thee. 8 And Solomon said unto God, Thou hast showed great mercy unto David my father, and hast made me to reign in his stead. 9 Now, O Lord God, let thy promise unto David my father be established: for thou hast made me king over a people like the dust of the earth in multitude. 10 Give me now wisdom and knowledge, that I may go out and come in before this people: for who can judge this thy people, that is so great? 11 And God said to Solomon, Because this was in thine heart, and thou hast not asked riches, wealth, or honour, nor the life of thine enemies, neither yet hast asked long life; but hast asked wisdom and knowledge for thyself, that thou mayest judge my people, over whom I have made thee king: 12 Wisdom and knowledge is granted unto thee; and I will give thee riches, and wealth, and honour, such as none of the kings have had that have been before thee, neither shall there any after thee have the like. (II Chronicles 1:7-12)*

In the first chapter of II Chronicles, the Hebrew word "mada" was used three times (in verses 10, 11, and 12). The word was translated as knowledge and spoke of what Solomon requested from God instead of effects like riches, long life, and victory over his enemies. God recognized that Solomon requested a cause instead of these effects, and God said He would give Solomon this cause as well as the effects.

The Hebrew word "mada" (H4093) comes from the root word "yada" (H3045) -"to know" or "to separate out mentally." It turns out mada means to know how you know and even got translated as the word "science" in Daniel 1:4. This word mada essentially means the ability to think about thinking. Let's see how it works.

In I Kings, the same story was told and the immediate application of this mada was seen.

> *Then there came two women, that were harlots, unto the king, and stood before him. And the one woman said, O my lord, I and this woman dwell in one house; and I was delivered of a child with her in the house. And it came to pass the third day after that I was delivered, that this woman was delivered also: and we were together; there was no stranger with us in the house, save we two in the house. And this woman's child died in the night: because she overlaid it. And she arose at midnight, and took my son from beside me, while thine handmaid slept, and laid it in her bosom, and laid her dead child in my bosom. And when I rose in the morning to give my child suck, behold, it was dead: but when I had considered it in the morning, behold, it was not my son, which I did bear. And the other woman said, Nay; but the living is my son, and the dead is thy son. And this said, No; but the dead is thy son, and the living is my son. Thus they spake before the king.* (I Kings 3:16-22)

This story is the ideal example to see the different ways Solomon could have dealt with this problem.

- **Absolve:** Ignore the women and send them away.
- **Resolve:** Punish or kill both women for bothering the king.
- **Solve:** Interview the women and their friends to determine who the mother is.

Notice, solving this problem is going to take a lot of effort, perhaps causing people to lie or blackmail each other, while matters requiring Solomon's attention would be ignored, with the possibility Solomon will still reach a wrong conclusion, which would be found out as the child got older and displayed a physical attribute similar to the actual mother. The woman who lost may have supporters who didn't agree with the answer and looked for ways to steal the child or kill the other woman. The child may even become traumatized, always wondering who their real mother was, as well as Solomon's handling of this matter opening the door for more of these types of disputes being brought to the king.

Think about it: Solve is our approach today through the courts. Do you think our legal process has made our nation better over time, or is it the best we can hope for to avoid lawlessness? Would a court's decision be the end of the matter, or would there be appeals? What was Solomon's approach, especially considering he had mada?

> *Then said the king, The one saith, This is my son that liveth, and thy son is the dead: and the other saith, Nay; but thy son is the dead, and my son is the living. And the king said, Bring me a sword. And they brought a sword before the king. And the king said, Divide the living child in two, and give half to the one, and half to the other. Then spake the woman whose the living child was unto the king, for her bowels yearned upon her son, and she said, O my lord, give her the living child, and in no wise slay it. But the other said, Let it be neither mine nor thine, divide it.*

> *Then the king answered and said, Give her the living child, and in no wise slay it: she is the mother thereof.* (I Kings 3:23-27)

Solomon took a step in the opposite direction of solve and analysis. Solomon said to give him a sword so he could divide the baby in two. What did that answer cause these two women to state?

They stated their thought processes; their *intangible* causes. Solomon was then able to know for sure which woman was the mother because he got to the intangible causes instead of looking for observable *whats*. Not only did Solomon's answer settle the matter forever without any appeals, it also made an interesting impression on all of Israel that was stated in the concluding verse of the chapter:

> *And all Israel heard of the judgment which the king had judged; and they feared the king: for they saw that the wisdom of God was in him, to do judgment.* (I Kings 5:28)

Basically, when people heard Solomon's decision, they thought, "Solomon is wise; however, that decision didn't come from a man. That decision could only have come from God." Hmmm, it looks like mada and dissolve are proofs of the wisdom of God! Let's look at another example.

Red Sea

The Israelites are in bondage to the Egyptians.

> *And it came to pass in process of time, that the king of Egypt died: and the children of Israel sighed by reason of the bondage, and they cried, and their cry came up unto God by reason of the bondage. And God heard their groaning, and God remembered*

his covenant with Abraham, with Isaac, and with Jacob. And God looked upon the children of Israel, and God had respect unto them. (Exodus 2:23-25)

What are God's possible responses?

- **Absolve:** Ignore the cries of the Israelites and maybe they will stop. (They won't.)
- **Resolve:** Punish the Egyptians until they freed the Israelites and/or allowed them to be Egyptian citizens.
- **Solve:** Come up with a plan that organizes the Israelites to the point they defeat the Egyptians in a revolt.

The first thirteen chapters of Exodus covered God's plan to deliver the Israelites out of Egypt. The process was explained in *Modeling God*: God acquired spiritual value through justice at Pharaoh's expense. Moses asked a question and Pharaoh considered the question an injustice he had to equal out. In reality, Pharaoh was doing an injustice and God was able to equal it out through a plague. Each plague caused Pharaoh to think he was justly able to punish the Israelites to a greater extent, when in reality, Pharaoh's responses allowed God to justly increase the severity of the next plague. Notice, the cause of this plan is intangible: Pharaoh's thought process relative to justice.

By the end of the thirteenth chapter of Exodus, the Israelites plundered the Egyptians and left because Pharaoh told them to leave in response to the last plague. Then what? Once Israel left, how far would they get before Pharaoh realized he had a trained army and the Israelites were a wandering hoard? Egypt would eventually send its army to attack the defenseless children of Israel. Exodus 14 began with:

> *And the Lord spake unto Moses, saying, Speak unto the children of Israel, that they turn and encamp before Pihahiroth, between Migdol and the sea, over against Baalzephon: before it shall ye encamp by the sea. For Pharaoh will say of the children of Israel, They are entangled in the land, the wilderness hath shut them in.* (Exodus 14:1-3)

God's plan addressed defending Israel from Egypt's army. However, rather than wait for later, God wanted to address it now. God's plan was to cause Egypt to see the children of Israel as confused and physically trapped by geography. Then God would save them with a supernatural event. It seems the intangible cause God addressed was Pharaoh's pride.

The parting of the Red Sea was not something Egypt would have considered in their prideful state. However, this meant circumstances had to appear to get worse for the children of Israel in order to lure in the Egyptian army. The end result: the Egyptian army was removed without the Israelites having to deal directly with them now or in the future. Problem dissolved!

Jesus

Our third example is documented in John chapter 8. Jesus was in the temple when the Pharisees brought a woman caught in the act of adultery. The Pharisees stated that the law said the woman ought to be stoned; however, they asked Jesus what He thought ought to be done. They trapped Jesus in a no-win situation. Either He had to encourage everyone to stone her and break Roman Law or tell everyone not to stone her and prove He wasn't keeping Moses' Law. Remember, the Jews were under Roman Law, so getting Jesus to break Roman Law would have gotten Him removed from society. Either way, the religious leaders thought they had solved their "Jesus problem."

Here were Jesus' options:

- **Absolve:** Ignore the situation, maybe it goes away. (The Pharisees continued to ask.)
- **Resolve:** Jesus could have tried to knock the stones out of everyone's hands so they couldn't stone her or tried to convince the crowd not to stone her, but this would have only energized the crowd and made Him look like He wasn't representing God.
- **Solve:** Jesus could have gotten into a doctrinal debate with the Pharisees, but He would have to admit Moses' Law wasn't wrong, and the woman would have been stoned. Jesus would have been the cause and arrested by the Romans. Worse, the Pharisees would have turned the crowd from Jesus if He contradicted Moses' Law or lost the debate with the Pharisees.
- **Dissolve:** Jesus came up with a way to stone her! It appears His response dealt with the intangible cause, the Pharisees' hypocrisy.

So when they continued asking him, he lifted up himself, and said unto them, He that is without sin among you, let him first cast a stone at her. And again he stooped down, and wrote on the ground. And they which heard it, being convicted by their own conscience, went out one by one, beginning at the eldest, even unto the last: and Jesus was left alone, and the woman standing in the midst. (John 8:7-9)

Jesus sat down after coming up with a way to stone her, and He (and the woman) didn't have to deal with the crowd. Jesus' statement was a conjunctive. It was truth, which we are beginning to see is mada.

Summary

When it comes to dealing with problems involving people, God's and Jesus' approach was not the analytical, solve approach that man naturally embraces, which focuses on the tangible causes. The man's will approach is making the world worse.

> *For as the heavens are higher than the earth, so are my ways higher than your ways, and my thoughts than your thoughts.* (Isaiah 55:9)

God's and Jesus' approach was the synthesis, dissolve approach that looked at the intangible spiritual causes. Realize that thought processes, definitions, and doctrine are all intangible spiritual causes.

Can you see how dissolve is in the God's will category? Dissolve is *how* God approaches everything that needs to be done to accomplish His plan because His plan involves people!

Can you begin to see more of the images on the wall? In the gold level we can see a bull's eye and the words: Leader, Spiritual, Qualitative, Wisdom, Dissolve, and Transformation. In the silver level, we can see an inner ring and the words: Understanding and Solve. In the brass level we can see the entire archery target and the words: Knowledge and Resolve. In the iron level we can see a wayward arrow and the words: Boss, Physical, Quantitative, Foolishness, and Absolve. We can also see God is the Husbandman and wants everyone to progress up to Him!

Dr. Joel Swokowski's Commentary

Jesus

There are so many examples to choose from with Jesus as He was always approaching problems with people through the dissolve method. I think of it along these lines: Whenever I read anything with Jesus in the story, He's walking around with the mindset of, "What's *really* going on here?" He sees the physical context the same as any man; yet, He's focused on what's going on behind the scenes.

- If a person is blind, He thinks, "What's *really* going on here?"
- If a person is causing a stir among the community, He thinks, "What's *really* going on here?"
- If a disciple asks who is the greatest among them, He thinks, "What's *really* going on here?"
- If a Samaritan woman asks for salvation, He thinks, "What's *really* going on here?"

Yes, even when a person asks Jesus for salvation, His first step wasn't to just say, "Great, repeat after Me...."

Let's dig into John 4:13-18 to see this interaction with the Samaritan woman.

> *Jesus answered and said unto her, Whosoever drinketh of this water shall thirst again: But whosoever drinketh of the water that I shall give him shall never thirst; but the water that I shall give him shall be in him a well of water springing up into everlasting life. The woman saith unto him, Sir, give me this water, that I thirst not, neither come hither to draw.* (John 4:13-15)

Jesus explained to her what He meant by living water, and His explanation intrigued her enough to respond by asking for that water. This was a great example of evangelism. Also, it makes me wonder how I would respond to this woman. I'd be tempted to quickly lead her in a prayer to receive Christ as her Savior. But this may have hurt her in the long term.

> *Jesus saith unto her, Go, call thy husband, and come hither.* (John 4:16)

Well, here you go! Jesus addressed something completely different than her request. Jesus saw and addressed a spiritual issue within this woman.

> *The woman answered and said, I have no husband. Jesus said unto her, Thou hast well said, I have no husband: For thou hast had five husbands; and he whom thou now hast is not thy husband: in that saidst thou truly.* (John 4:17-18)

Since Jesus addressed the spiritual issue, we really don't know what would have happened if Jesus had done anything other than this. Yet, I presume that if He just led her into salvation without addressing the husband issue, the husband issue would have still been there after she became born again. This issue may have caused her to be distracted from this new spiritual life and resulted in her forfeiting her salvation.

Instead, Jesus first addressed the very thing that would have resulted in her long-term destruction. In fact, her being with a man who wasn't her husband means she would have been in adultery. Would it have been right and just of Jesus to ignore this and focus on getting someone to say a salvation prayer so He could be on His way? It looks like this was the second documented incident of how Jesus interacted with an actively adulterous woman.

If you read the rest of the story, this caused the woman to ask Jesus a doctrinal question, and His answer caused her not to pursue salvation in the moment but to go into her city and tell people about her encounter. The result?

> *So when the Samaritans were come unto him, they besought him that he would tarry with them: and he abode there two days. And many more believed because of his own word; And said unto the woman, Now we believe, not because of thy saying: for we have heard him ourselves, and know that this is indeed the Christ, the Saviour of the world.* (John 4:40-42)

It looks like this was the perfect will of God, and my solve or resolve approach may not have even qualified as God's good will in the long term. The perfect will of God is always to dissolve.

CHAPTER 13

God's Four Measures for Judgment

WHY DID GOD judge Sodom? If you say because of immoral sexual behavior, those are the effects, not the causes of judgment from God. Remember, God wants us to understand His will: *why* and *how* He moves for or against an individual or group. As always, the only place to get the answer for sure is from the Bible.

> *Behold, this was the iniquity of thy sister Sodom, pride, fulness of bread, and abundance of idleness was in her and in her daughters, neither did she strengthen the hand of the poor and needy.* (Ezekiel 16:49)

The four causes of judgment against Sodom were:

1. Pride
2. Fullness of bread
3. Abundance of idleness (also interpreted as "prosperous ease")
4. Not strengthening the poor and needy

Notice, God blatantly stated the four causes of judgment against Sodom. Anything else stated as a cause of the judgment of Sodom is not in line with the word of God.

Pride is arrogance; when an individual or group of people sees himself or herself as better than everyone else. It is the opposite of humility which is being able to consider you are wrong. Pride is shown when the person can't see, nor state, any way they could possibly be wrong. It is the opposite of contrastive thinking. Ultimately, it leads to the belief by the individual that he is equal to or greater than God and is shown when someone thinks their good fortune is due more to their ability to create than God's ability.

Fullness of bread is not just overeating, it also applies to a focus on the flesh and continually filling up the senses with tangible things past what is necessary. This treats the physical as the cause. This is a graphic example of consumption and waste. With "fullness of bread," resources are literally being consumed with no increase in return. In fact, the return actually diminishes due to poor health, shortened life, and increased medical expenses. Also, in terms of helping others, think about what could be accomplished with the resources.

Likewise, abundance of idleness or prosperous ease means people aren't using their time wisely. This is comfort instead of growth. How much more could people be helped if we spent our time wisely? Clearly, these first three causes result in unprofitability, the opposite of God's measure for growth.

The final cause (not strengthening the poor and needy) is people not helping others who are in need. Like the previous three, this trait is wasteful. However, since people are eternal, this is an act of spiritual consumption. This is contradictory to having the Spirit of God in you, which means it is an act of holding back the Holy Spirit from being profitable.

> *But whoso hath this world's goods, and seeth his brother have need, and shutteth up his bowels of compassion from him, how dwelleth the love of God in him?* (I John 3:17)

I understand this can seem theoretical; however, we will see these four measures consistently used throughout the Bible as a characteristic of God's will. We can see why these four causes bring judgment when we view them against the four causes of Modeletics: contrastive thinking, causality, growth-mindset, and non-contradiction. In order to make this clearer, let's take an in-depth look at someone in the Bible who is very misunderstood because people don't know these measures.

Solomon

We saw Solomon had mada in response to wanting to rule others justly. In fact, Solomon was the living example of what most people think will make them happy: wealth, wisdom, power, and no enemies. However, did you know that Solomon ended his life as a failure by worshiping other gods without repentance? Let's do a deep dive to see what happened.

We saw that Moses changed God's will when He wanted to wipe out the Israelites and start over with Moses. We also saw God was eventually proven right. In Deuteronomy 17, we see Moses realized God was right. He told the people they would eventually want a king like the other nations and gave them the following commands:

> *When thou art come unto the land which the Lord thy God giveth thee, and shalt possess it, and shalt dwell therein, and shalt say,*

> *I will set a king over me, like as all the nations that are about me; Thou shalt in any wise set him king over thee, whom the Lord thy God shall choose: one from among thy brethren shalt thou set king over thee: thou mayest not set a stranger over thee, which is not thy brother. But he shall not multiply horses to himself, nor cause the people to return to Egypt, to the end that he should multiply horses: forasmuch as the Lord hath said unto you, Ye shall henceforth return no more that way. Neither shall he multiply wives to himself, that his heart turn not away: neither shall he greatly multiply to himself silver and gold. And it shall be, when he sitteth upon the throne of his kingdom, that he shall write him a copy of this law in a book out of that which is before the priests the Levites: And it shall be with him, and he shall read therein all the days of his life: that he may learn to fear the Lord his God, to keep all the words of this law and these statutes, to do them: That his heart be not lifted up above his brethren, and that he turn not aside from the commandment, to the right hand, or to the left: to the end that he may prolong his days in his kingdom, he, and his children, in the midst of Israel.* (Deuteronomy 17:14-20)

Did Solomon do this when he became king? Here was how I Kings 10 ended and I Kings 11 began:

> *And Solomon gathered together chariots and horsemen: and he had a thousand and four hundred chariots, and twelve thousand horsemen, whom he bestowed in the cities for chariots, and with the king at Jerusalem. And the king made silver to be in Jerusalem as stones, and cedars made he to be as the sycomore trees that are in the vale, for abundance. And Solomon had horses brought out of Egypt, and linen yarn: the king's merchants received the linen yarn at a price. And a chariot came up and went out*

> *of Egypt for six hundred shekels of silver, and an horse for an hundred and fifty: and so for all the kings of the Hittites, and for the kings of Syria, did they bring them out by their means.* (I Kings 10:26-29)

> *But king Solomon loved many strange women, together with the daughter of Pharaoh, women of the Moabites, Ammonites, Edomites, Zidonians, and Hittites: Of the nations concerning which the Lord said unto the children of Israel, Ye shall not go in to them, neither shall they come in unto you: for surely they will turn away your heart after their gods: Solomon clave unto these in love. And he had seven hundred wives, princesses, and three hundred concubines: and his wives turned away his heart. For it came to pass, when Solomon was old, that his wives turned away his heart after other gods: and his heart was not perfect with the Lord his God, as was the heart of David his father.* (I Kings 11:1-4)

Clearly, Solomon didn't read God's law all the days of his life, otherwise he would have realized he was violating the commands given in Deuteronomy 17!

Despite having all of these tangible blessings, Solomon was unhappy. The proof of Solomon's unhappiness was shown in the Book of Ecclesiastes, which Solomon wrote at the end of his life. Furthermore, it is the ultimate example of how even the wisest man who ever lived could violate these four areas. Let's start with the first two verses of the book.

> *The words of the Preacher, the son of David, king in Jerusalem. Vanity of vanities, saith the Preacher; vanity of vanities, all is vanity.* (Ecclesiastes 1:1-2)

Here were the last two verses of the first chapter:

> *And I gave my heart to know wisdom, and to know madness and folly: I perceived that this also is vexation of spirit. For in much wisdom is much grief: and he that increaseth knowledge increaseth sorrow.* (Ecclesiastes 1:17-18)

The first chapter is an example of how Solomon repeatedly stated everything was pointless and unprofitable to try to understand. He justified idleness, which was a mental issue.

Next, if you look at Ecclesiastes Chapter 2, you'll see Solomon used the entire chapter attempting to prove he had more than anyone and knew more than anyone. This was pride. Pride is an emotional issue. It really is trying to feel better because you see yourself as better than everyone else. Next, look at Ecclesiastes 2:24.

> *There is nothing better for a man, than that he should eat and drink, and that he should make his soul enjoy good in his labour. This also I saw, that it was from the hand of God.*

Looking to the fullness of bread for happiness is a physical issue. Solomon repeated this conclusion several times. Worse, Solomon's labor was building high places to other gods, likely to pacify his wives or maybe even to get blessings from those gods! For the final cause of judgment, check out Ecclesiastes 5:8-9.

> *If thou seest the oppression of the poor, and violent perverting of judgment and justice in a province, marvel not at the matter: for he that is higher than the highest regardeth; and there be higher than they. Moreover the profit of the earth is for all: the king himself is served by the field.*

Solomon actually justified that it was not his job to help the poor! As we saw above, not strengthening the poor and needy is a spiritual issue. Here was how Solomon ended the book:

> *Let us hear the conclusion of the whole matter: Fear God, and keep his commandments: for this is the whole duty of man. For God shall bring every work into judgment, with every secret thing, whether it be good, or whether it be evil.* (Ecclesiastes 12:13-14)

Solomon ended Ecclesiastes stating he knew what to do and he knew no one could hide anything from God. Many people think this meant Solomon did fear (respect) God and kept His commandments; however, we already showed that he didn't. In fact, God told Solomon to take down the high places to other gods, and we see he never did it, as evidenced by Josiah taking down Solomon's high places hundreds of years later:

> *And the high places that were before Jerusalem, which were on the right hand of the mount of corruption, which Solomon the king of Israel had builded for Ashtoreth the abomination of the Zidonians, and for Chemosh the abomination of the Moabites, and for Milcom the abomination of the children of Ammon, did the king defile. And he brake in pieces the images, and cut down the groves, and filled their places with the bones of men.* (II Kings 23:13-14)

How could Solomon have fixed this? The irony is when Solomon dedicated the temple he built to God, he had the following exchange with God:

> *Thus Solomon finished the house of the Lord, and the king's house: and all that came into Solomon's heart to make in the house of the Lord, and in his own house, he prosperously effected. And the*

Lord appeared to Solomon by night, and said unto him, I have heard thy prayer, and have chosen this place to myself for an house of sacrifice. If I shut up heaven that there be no rain, or if I command the locusts to devour the land, or if I send pestilence among my people; If my people, which are called by my name, shall humble themselves, and pray, and seek my face, and turn from their wicked ways; then will I hear from heaven, and will forgive their sin, and will heal their land. (II Chronicles 7:11-14)

Notice, God was undoing the four causes of judgment with humility, prayer, spending your time seeking God's face, and turning from your wicked ways. God had told Solomon what to do, and Solomon chose not to do it. Worse, it looks as if Solomon did his will for decades to the point it seared his conscience, and even though he could state what he ought to do, he couldn't actually get the energy to do it.

What are examples of how to do the opposite of the four causes of judgment so you receive blessings? Let's look at an example for an individual and an example for a group.

Blessed Examples

Obviously, the ultimate individual example was Jesus.

1. Pride: Jesus responded with humility.

Let this mind be in you, which was also in Christ Jesus: Who, being in the form of God, thought it not robbery to be equal with God: But made himself of no reputation, and took upon him the form of a servant, and was made in the likeness of men: And being found in fashion as a man, he humbled himself, and became obedient unto death, even the death of the cross. (Philippians 2:5-8)

2. Fullness of bread: Jesus responded with fasting.

And when he had fasted forty days and forty nights, he was afterward an hungered. (Matthew 4:2)

3. Abundance of idleness (prosperous ease): Jesus responded by doing God's works.

I must work the works of him that sent me, while it is day: the night cometh, when no man can work. (John 9:4)

4. Not strengthening the poor and needy: Jesus responded by strengthening them.

For even the Son of man came not to be ministered unto, but to minister, and to give his life a ransom for many. (Mark 10:45)

And he lifted up his eyes on his disciples, and said, Blessed be ye poor: for yours is the kingdom of God. (Luke 6:20)

The group example was what people refer to as "the Acts church." It was the church that formed immediately after Pentecost and was described in Acts 2.

And they continued stedfastly in the apostles' doctrine and fellowship, and in breaking of bread, and in prayers. And fear came upon every soul: and many wonders and signs were done by the apostles. And all that believed were together, and had all things common; And sold their possessions and goods, and parted them to all men, as every man had need. And they, continuing daily with one accord in the temple, and breaking bread from house to house, did eat their meat with gladness and singleness of heart, Praising God, and having favour with all the people. And the Lord added to the church daily such as should be saved. (Acts 2:42-47)

Here was how the four causes of church lined up with the four causes of judgment:

1. **Pride:** fellowship
2. **Fullness of bread:** breaking of bread, sharing
3. **Abundance of idleness (Prosperous Ease):** apostle's doctrine
4. **Not strengthening the poor and needy:** helping all that had need, prayer (spiritual issue)

The four causes of church directly addressed the four causes of judgment on Sodom.

Summary

God looks at the whole person: mentally, emotionally, physically, and spiritually. Likewise, He looks at groups of people the same way. This is why God can judge a group, like Israel and Judah, as He judges a person because He's judged the whole person. We called this synthesis systems thinking.

Man's will is characterized by analysis and the four causes of judgment: pride, fullness of bread, abundance of idleness, and not strengthening the poor and needy.

Do you see how this topic is in line with God's will: *how* He accomplishes His plan?

We learned more about *how* God determines who is on the wrong side of justice by looking at four measures that allow God to justly bring judgment on people and nations, regardless of whether they say they believe in God or not.

How is the country you live in doing with these four causes of judgment?

Dr. Joel Swokowski's Commentary

Pride

Due to the nature of the sin of pride and its implications, I want to bring some clarity to this term. The Hebrew word translated into the word pride means exaltation, elevation, etc. (H1347 gaon). C.S. Lewis considered pride to be the one vice that holds true for every person. He referred to it as self-conceit, which is directly in line with the definition.

Simply put, pride is the belief that you are superior.

The effects of that are a person who is unteachable, believes they are in need of nothing, and have no ability to consider a perspective outside of their own including that they could be wrong.

Understanding the definition of pride as a belief that you are superior leads to greater clarity for this famous Bible verse:

> *Pride goeth before destruction, and an haughty spirit before a fall.* (Proverbs 16:18)

A person in pride cannot be taught anything, and believing that no one can teach you causes you to believe you don't need to change, which eventually results in error (and ultimately destruction) because none of us are perfect.

Notice that pride and humility are contrasted when it comes to interacting with God:

> *But he giveth more grace. Wherefore he saith, God resisteth the proud, but giveth grace unto the humble.* (James 4:6)

The Book of Proverbs consistently says that fear of God (respect for Jehovah) and humility are the causes for the desired effects.

> *By humility and the fear of the Lord are riches, and honour, and life.* (Proverbs 22:4)

Modeletics and the Four Causes of Judgment

One of the incredible correlations to the information in this chapter is how the causes of church dissolve the causes for judgment. Lenhart wrote the following above:

Here was how the four causes of church lined up with the four causes of judgment:

1. ***Pride:*** *fellowship*
2. ***Fullness of bread:*** *breaking of bread, sharing*
3. ***Abundance of idleness (Prosperous Ease):*** *apostle's doctrine*
4. ***Not strengthening the poor and needy:*** *helping all that had need, prayer (spiritual issue)*

The four causes of church directly addressed the four causes of judgment on Sodom.

Furthermore, the four causes of church can be seen as fulfilling every need that a person (or group of people) may have. Think about it, if you and your family had fellowship, food, doctrine, and answered prayers, what more would you need? These four causes sustain us indefinitely. We've also seen that these four causes of judgment and church correlate to the emotional, physical, mental, and spiritual parts of humanity. There is no need left unfilled, and no part of humanity left unsatisfied.

Lenhart also stated these four causes of judgment lined up with the four causes of Modeletics. Here is how the four causes of Modeletics would address the four causes of judgment:

1. **Pride:** contrastive thinking
2. **Fullness of bread:** causality (The physical is not the cause.)
3. **Abundance of idleness (prosperous ease):** growth-mindset vs. comfort
4. **Not strengthening the poor and needy:** non-contradiction (Each of us looks to others when we are in need and not doing this is contradictory to the belief one has the love of God.)

If Solomon had used the principles of Modeletics, which are God-given truths found in His word, the Book of Ecclesiastes would have had a much different message, or not have been written at all. It is the holding of a growth-mindset that facilitates a person into embracing causality and using contrastive thinking in order to remove contradictions that would have led Solomon (and, for that matter, Sodom) into living a life in pursuit of the spiritual and intangible over the physical and tangible stuff he accrued.

Finally, we can see the commands God gave to kings (that Solomon violated) as addressing each part of a person and possibly correlating with the four causes of judgment.

1. **Emotional:** Solomon's pride in violating the three commands of God and possibly not making his own copy of the Torah and reading it.
2. **Physical:** making silver plentiful so people could have fullness of bread.
3. **Mental:** horses to distract people because they had an abundance of idleness.
4. **Spiritual:** multiplying wives and concubines led Solomon's heart away from God and perhaps hampered Solomon's ability to care for the poor and needy because he had 1000 women and their kids to take care of.

CHAPTER 14

Prophecy

MOST PEOPLE THINK prophecy proves that the future has already happened, which is determinism/Calvinism. The purpose of prophecy is to share and confirm God's will and God's plan, which ought to increase our faith. God's true prophets revealed His will to the people.

> *Surely the Lord God will do nothing, but he revealeth his secret unto his servants the prophets.* (Amos 3:7)

We saw with Ezekiel that God said He was sharing His will ahead of time so it would comfort people and build their faith because they would know it was God. While it is only right and just for God to share His will so we can make sure we are not fighting the will of God, people still see prophecy as completely focused on the *what* with no regard given to the *why* and *how*, and it's mostly because of passages like this:

> *And if thou say in thine heart, How shall we know the word which the Lord hath not spoken? When a prophet speaketh in the name of the Lord, if the thing follow not, nor come to pass, that is the thing which the Lord hath not spoken, but the prophet hath spoken it presumptuously: thou shalt not be afraid of him.* (Deuteronomy 18:21-22)

Keeping this in mind, take another look at the story of Hezekiah that we covered previously.

> *In those days was Hezekiah sick unto death. And the prophet Isaiah the son of Amoz came to him, and said unto him, Thus saith the Lord, Set thine house in order; for thou shalt die, and not live.*
>
> *Then he turned his face to the wall, and prayed unto the Lord, saying, I beseech thee, O Lord, remember now how I have walked before thee in truth and with a perfect heart, and have done that which is good in thy sight. And Hezekiah wept sore.*
>
> *And it came to pass, afore Isaiah was gone out into the middle court, that the word of the Lord came to him, saying, Turn again, and tell Hezekiah the captain of my people, Thus saith the Lord, the God of David thy father, I have heard thy prayer, I have seen thy tears: behold, I will heal thee: on the third day thou shalt go up unto the house of the Lord. And I will add unto thy days fifteen years; and I will deliver thee and this city out of the hand of the king of Assyria; and I will defend this city for mine own sake, and for my servant David's sake.* (II Kings 20:1-6)

Isaiah stated that Jehovah said Hezekiah would die now, but Hezekiah died fifteen years later. Isaiah stated *what* would happen and it didn't happen. Was Isaiah a prophet? Jesus said:

> *Ye hypocrites, well did Esaias prophesy of you, saying* (Matthew 15:7)

Jesus referenced Isaiah (Esaias) as a prophet, so he was a prophet. Therefore, the only explanation that is non-contradictory with the rest of

God's word is: Hezekiah's prayer and his righteousness (reward) caused God to move and heal Hezekiah, and God did not know everything that would happen ahead of time.

God knows all the causes that exist and their effects. God does not know the causes that do not yet exist. At the beginning of this story, the cause that existed was Hezekiah's sickness which would result in his death and Isaiah stated this result, that is, *what* would happen. Hezekiah's prayer added a new cause to the story, which changed the result; it changed the *what*. What do you think is God's reason for prophecy?

Theologians can't give a non-contradictory explanation for this story because they want to believe God had everything figured out ahead of time, or they don't understand the doctrines of prayer and prophecy. Either God lied, or God didn't have everything exactly figured out ahead of time. Theologians, which one is it?

Prophecy is a statement of the effects that will happen if current causes stay the same. The reason prophecy is given is for God to share His will at that moment in time. Let's look at another case.

Jonah

Jesus called Jonah (Jonas) a prophet.

> *But he answered and said unto them, An evil and adulterous generation seeketh after a sign; and there shall no sign be given to it, but the sign of the prophet Jonas:* (Matthew 12:39)

We have seen how Jonah was in God's good will. Let's pick up the story in Jonah chapter 3 after God stressed Jonah to the point he agreed to obey God's will.

> *And the word of the Lord came unto Jonah the second time, saying, Arise, go unto Nineveh, that great city, and preach unto it the preaching that I bid thee.* (Jonah 3:1-2)

It looks as if God was beginning the process anew once the fish vomited Jonah on the shore. God told Jonah to go to Nineveh and preach the message that God commanded of him.

> *So Jonah arose, and went unto Nineveh, according to the word of the Lord. Now Nineveh was an exceeding great city of three days' journey. And Jonah began to enter into the city a day's journey, and he cried, and said, Yet forty days, and Nineveh shall be overthrown.* (Jonah 3:3-4)

Jonah stated that in forty days, this city will be overthrown. Jonah didn't state *might be*, but rather *will be* overthrown. This is what Jonah stated from Jehovah.

> *So the people of Nineveh believed God, and proclaimed a fast, and put on sackcloth, from the greatest of them even to the least of them. For word came unto the king of Nineveh, and he arose from his throne, and he laid his robe from him, and covered him with sackcloth, and sat in ashes. And he caused it to be proclaimed and published through Nineveh by the decree of the king and his nobles, saying, Let neither man nor beast, herd nor flock, taste any thing: let them not feed, nor drink water: But let man and beast be covered with sackcloth, and cry mightily unto God: yea, let them turn every one from his evil way, and*

> *from the violence that is in their hands. Who can tell if God will turn and repent, and turn away from his fierce anger, that we perish not?* (Jonah 3:5-9)

The Ninevites believed God would do this! What was their response? Was it to run for the hills or spend their last forty days in debauchery? No, it was to confess and repent! Furthermore, it appeared that Nineveh's response was to do the opposite of the causes of Sodom's judgment!

1. **Pride:** They humbled themselves.
2. **Fullness of bread:** They fasted.
3. **Abundance of idleness:** They turned from violence.
4. **Strengthen the arm of the poor and needy:** They prayed (spiritual cause), perhaps confessing their sin in not strengthening the arm of the poor and needy.

What was God's response?

> *And God saw their works, that they turned from their evil way; and God repented of the evil, that he had said that he would do unto them; and he did it not.* (Jonah 3:10)

Why did the king choose to do these four things? Clearly, Nineveh was doing the same four causes of judgment against Sodom. Then, after Jonah's prophecy, they changed their behavior to doing the opposite of the four causes of judgment. Is it possible the king knew the causes of God's judgment against Sodom? Whether he knew them or not, the king's response resulted in God turning (repenting) from the evil (destruction) that He was going to do to them. The passage specifically said, "*and he did it not.*"

Jesus called Jonah a prophet, and what Jonah said would happen did *not* happen. Was Jesus endorsing false prophets? Again, prophecy is a statement of the effects that will happen if current causes stay the same. Therefore, the greatest value of prophecy to people is for people to understand God's will so people are given time to repent and intentionally express their will.

Simple vs. Complex

Notice, prophecy is God stating what He will do when it is His turn to respond after He has enough spiritual value through justice. When it comes to prophecy, there seems to be two types because God's ability to respond is according to one of two situations. What we've seen in this chapter is simple prophecy. God was stating the result He would respond with based on the causes that currently existed.

In order to understand the other type of prophecy, let's look at the seventy weeks of Daniel that was mentioned in a previous chapter (with commentary):

> *Seventy weeks are determined upon thy people and upon thy holy city, to finish the transgression, and to make an end of sins, and to make reconciliation for iniquity, and to bring in everlasting righteousness, and to seal up the vision and prophecy, and to anoint the most Holy.* (Daniel 9:24)

This was a simple prophecy where a week refers to a seven-year period. God was declaring that the people of Israel and Jerusalem will have 490 years to bring an end to sin and bring in everlasting righteousness with the anointing of the most Holy, which would also bring prophecy to an end. Daniel got a prophecy about the end of prophecy! What could be simpler than that? Let's continue with the passage.

> *Know therefore and understand, that from the going forth of the commandment to restore and to build Jerusalem unto the Messiah the Prince shall be seven weeks, and threescore and two weeks: the street shall be built again, and the wall, even in troublous times. And after threescore and two weeks shall Messiah be cut off, but not for himself: and the people of the prince that shall come shall destroy the city and the sanctuary; and the end thereof shall be with a flood, and unto the end of the war desolations are determined.* (Daniel 9:25-26)

This 490-year period would begin when the commandment is given to restore and build Jerusalem. When had that occurred? It hadn't yet. This was what makes this prophecy complex. God was stating an event that would begin or trigger the simple prophecy.

Why couldn't God state when that trigger event would happen? Why could God state a specific timeline after that trigger event? Let's continue because God only accounted for 483 years of the 490-year timeline.

> *And he shall confirm the covenant with many for one week: and in the midst of the week he shall cause the sacrifice and the oblation to cease, and for the overspreading of abominations he shall make it desolate, even until the consummation, and that determined shall be poured upon the desolate.* (Daniel 9:27)

This final seven-year period is known as the Tribulation.

All the following were revealed in the Book of Revelation:

- The Seals
- The Bowls
- The Tribulation

- The Raptures
- The Millennial Reign
- The White Throne Judgment
- The Wedding Supper Of The Lamb

These are all future events that God told John about—things that will happen once God has enough value to respond.

We know that when this seven-year period begins, God will have enough value to respond and, clearly, He was sharing His response ahead of time to build the faith of the people living during that time that God caused this to happen.

Isn't that the answer for our two questions? In fact, not only does it sound like God would have enough spiritual value to share what He was going to do once the command to build the temple was given, it sounded like He would cause the command to build the temple *when* He had enough spiritual value. God could see the causes in place were going to result in Him eventually having enough spiritual value. He couldn't say exactly when because people affect the timing; however, He could state what He would do in response once He had enough spiritual value.

This type of prophecy tended to be associated with changes in dispensations. This passage showed The Tribulation ("one week") is part of the fifth dispensation! Remember, we saw that the fifth dispensation comes back after our current dispensation of grace is over.

This prophecy also showed that God gave Israel an objective warning when their dispensation would end and they would cut off the Messiah. No one had an excuse for not recognizing the Messiah and the change in dispensation, if they followed the word of God.

Summary

There are two types of prophecy. The first may or may not involve God taking an active role. He knows the causes, and the prophecy is His statement of the effects, whether He is going to bring them about or they are going to happen naturally. The second involves God. He knows the causes and the effects will cause Him to have enough spiritual value through justice that He will be able to respond. In this case, He told us ahead of time what His decision would be when He did eventually get enough spiritual value.

Now we see prophecy is really about God helping us get on the right path. God's goal isn't to see if He can predict the future. God uses prophecy to correct believers or bring confirmation, both of which build our faith that we are in God's will. Confirmation is an encouragement you are on the right path, while correction is meant to help you become profitable.

God values you being a part of His plan to the point He is concerned with every individual's growth. God wants to bring about His will by having His Righteousness work through you so that you receive salvation and reward.

Notice, it doesn't matter whether you are currently on the right path or not. What matters is your response to God's help in getting you on the right path.

Do people really want to grow, understand God more, be led by mada, and know the non-contradictory truth? Do they want to have the right causes? Do they want to have the right doctrine? These are characteristics of God's will.

Or do they want to be comfortable, comparative, attempt to keep God a mystery, and validate their personal contradictory beliefs that can seem right in the moment? Do they want to justify man-made doctrine? These are characteristics of man's will.

It looks as if the word "Mental" has been added to the silver level, and the word "Emotional" has been added to the brass level. The four principles of Modeletics are in the gold level with the word "Truth" and an image of a council meeting, while the four causes of judgment are in the iron level along with an image of Solomon.

We are going to conclude Book 1 by bringing everything together with a revelatory conclusion concerning how God brings about His will.

Dr. Joel Swokowski's Commentary

Trespass

The etymology of the word trespass is "cross beyond the path" (Oxford Languages). Prophecy is a technique God uses to keep us on the path (His will) or to get us back on the path. God is right and just, and the path He has set before every individual and group requires righteousness and justice for the person(s) to stay on the path. We're human, we make mistakes. We're not robots where you can just put a program in our brains and expect a perfect result (behavior) to come out every time.

God is right and just, and He knows we aren't. God is not expecting us to be right and just. He shows us mercy when we aren't right and just in the hopes of a good response from us. We are bound to drift from the path, no matter our intentions. The more we take direction from God, the more likely we'll stay on the path. The more we embrace wise counsel and mada, the more likely we'll stay on the path.

I see it like God managing our walk with little nudges. As I walk a straight line, I may start veering off to the left. At this point, even if I'm still on the path but close to taking a step off (trespass), God will give me a little nudge to put me in the center once again. Over time, I may start veering off to the right. At this point, even if I'm still on the path but close to taking a step off (trespass), God will give me a little nudge to put me in the center once again.

This can look like what I referred to in Chapter 5's commentary on "Adding to God's Word." If my church sings three songs of worship every Sunday service and it results in a supernatural experience for everyone in attendance, it could be tempting to add a fourth song. The intentions may even be good, "if three songs are good, four would be even better!" However, if God determined the three-song standard and I was determined to add the fourth, even if my intentions were good, it is an example of me adding to the word or command of God. It could be seen as me taking one step off the path (trespass).

If this method of operating continued, worship could eventually get to a place where it's all about the people, and not about God. Worse, this method of operating would eventually leak into other (all) areas of managing the church. Before you know it, one step off the path results in a church that is building man's name, not God's. So, when God nudges me in any area, even if it seems insignificant at the time, it may be God saving me from making huge and lasting adjustments that hurt people. I ought to be grateful for every nudge God sends my way!

CHAPTER 15

How God Brings About His Will

THIS FIRST BOOK has focused on explaining God's will. Here is a brief summary.

God has a plan for eternity. This is what God wants. We call it the party in our analogy and it is the ultimate in intimacy and communication by exchanging with others through our AREs (who God created us to be). How God brings about this plan is God's will. Likewise, God has a plan for individuals and groups, which means God has an individual will and a group will.

Not only does God have three wills, but God also has three levels within each will. Perfect is when God's will is immediately and completely followed. Pleasing or acceptable is when God's will is not immediately followed because the individual or group needs more faith, usually supplied by a leader who has a share in what is accomplished when they do follow God's will. Finally, good is when God increases the tension, and His will is chosen over complete disobedience. Complete disobedience is out of God's will.

We have seen that everything in the universe is a system except for God. A system is a whole made up of two or more essential parts, and God doesn't have any parts. God is Holy, of one substance. In fact, He is the only thing in the universe that is of one substance, with everything else being a system made up of parts. This proves He is the First Cause, the Creator. God's universal will is the greatest and most complicated system that exists and, consequently, is the main focus of this edition. Those who are able to determine God's will are said to be transformed.

We have seen that God brings about His universal will through people who have enough spiritual value (reward). Salvation is achieved by grace through faith, while reward is achieved by justice through our interactions with other people. Either we gain at someone's expense by handling the injustice well or we gain by exchanging with others through our uniquenesses.

The fact that God's universal will is dependent on reward is just one of the reasons this doctrine is very difficult for people to grasp. Most "Christians" are only focused on salvation, and many believe we either shouldn't be focused on reward or that everyone will get the same reward in heaven.

Remember, all of the parables Jesus told where people got the same reward were salvation parables. Everyone gets the same salvation. However, the parables where Jesus mentioned people getting different amounts of reward were reward parables. God is right and just. Would it be just of God for everyone to get the same reward considering some people use their reward to have physical pleasures now, or drain it through taking out their own justice on others, while others obey Jesus' words and store up this reward in heaven?

God's Story

If we are going to look at God's universal will, we need to understand God's story. The Bible documented God's story, and it is fairly easy to understand when the Bible is arranged chronologically, and one has the doctrine that is shared in these editions. Let's look at an overview of God's story in terms of characters, conflict, and resolution.

Characters: God, the other spiritual beings, and humanity.

Before we cover the conflict, we need to review God's goal. God wants to dwell for eternity in a kingdom made up of spiritual and physical beings. God wants to interact with these beings in eternity because it is the only way to achieve happiness and joy. Specifically, He, Jesus, and the Holy Spirit will interact with each other through the Bride, which is made up of willing humans acting in their uniqueness. Now, let's look at the conflict, or what is hindering this goal.

Conflict: God is only able to achieve this goal with and through people who are Righteous, and people do not want to be Righteous! Remember, it only takes one person to make the Bride unprofitable. Worse, throughout the Bible, when people get in groups, they consistently sink to the lowest thought process and reject God, the Person looking out for them the most!

Resolution: We're still living it! There will be a Judgment Day that objectively determines who is Righteous and who is not. The resolution for us humans is that we all choose our final destination: new Jerusalem or lake of fire! However, our focus right now is God's resolution to God's story.

Since God is right and just, He can't unilaterally resolve this conflict. God can only make His moves in response to right and just. Ultimately,

God can only make His moves when He has enough spiritual value. God's will is how God obtains the spiritual value to bring about the resolution to His story.

Since we have a role in God's story, let's first look more closely at our resolution because it will help us better explain God's resolution. What is our story?

Our Story

Characters: You, God, other people.

What is your goal? Everyone's goal is to be happy; to get their brain to release pleasure chemicals. Everyone does everything either to get their brain to release pleasure chemicals now or to set their brain up to release pleasure chemicals in the future. No one does anything that prevents their brain from releasing pleasure chemicals now and in the future.

> *But as many as received him, to them gave he power to become the sons of God, even to them that believe on his name: Which were born, not of blood, nor of the will of the flesh, nor of the will of man, but of God.* (John 1:12-13)

This verse proved, through a contrastive process, that the believers had to be born of God because it wasn't any of the other possible options. Notice, this verse told us there were not only four possible ways for an individual to get their brain to release pleasure chemicals, it showed there are only four thought processes your brain can operate through!

Before we look more closely at each one and how it would apply to bringing about one's goal of being happy, I have a question for you to answer. You don't have to share your answer with anyone, and you can tear up the paper after you have written it. However, please be honest with your answer because your unconscious brain and the Holy Spirit will see your answer, and if what you write isn't consistent with how you live your life, you could be bringing judgment on yourself.

Please write down your answer to this question: What do you need to have or do in order to be happy? (Notice: if you write down "value family" but you consistently choose work over your family, that answer could bring judgment on yourself. Again, for your own sake, be honest with your answer.)

Now that you have identified YOUR plan, let's continue!

Blood (Evil)

We know evil means to destroy. People who try to be happy through destruction are people who take pleasure in the misfortune of others. These people would answer the question above with: "Have my enemy die" or "Have my competition go out of business."

We all have this thought process in areas of our lives where we feel out of control. For example, when I ask people to close their eyes and imagine their enemy slipping on a banana peel, I have found that half of people laugh. Again, there is a difference between having this thought process guiding one aspect of your life versus it being the answer to your overall happiness.

Now we see that everyone's goal is to be happy; however, the issue is the *plan* the person chooses to be happy. Hitler was trying to be happy. What made Hitler evil was his *plan* for being happy. I have found the first step I take in my brain when I begin to deal with someone is to realize the person is trying to be happy, and then I see the plan they have chosen to be happy. Let's look at the next plan!

Flesh (Animal)

We know the flesh tries to be comfortable in the moment. People who try to be happy by avoiding tension are people who look for tangible things to bring them pleasure. These people would answer the question above with: "win the lottery, play video games/sports, retire, hobbies," and even "spend all day sitting under a cabana drinking mai tai's."

We all have this thought process in areas of our lives where we want to avoid tension, usually by distracting ourselves. If we saw the evil level as unprofitable, we could see this level as not unprofitable. After all, we aren't intentionally destroying value, we just aren't producing anything of lasting value.

Consequently, this thought process was also known in the Bible as the animal thought process. The following passages also refer to this as a beast or fleshly thought process: Jude 1:10, Psalm 49:20, Jeremiah 17:5, II Peter 2:12-13. In addition, most of us know this as the addictive or habitual thought process. Isn't this the thought process Solomon was in at the end of his life? He consciously knew what to do, but like an addict, he couldn't bring himself to actually do it.

The reality is your brain is made to be addicted to something. Isn't that what a habit is? Basically, you have trained your brain to feel uncomfortable when you don't do what you've always done and only feel satisfied when you do the habit.

Notice, the "happiness" felt by these people isn't a lasting feeling of fulfillment. This "happiness" is a reprieve from a bad feeling (boredom or tension), which in contrast is positive, and people train their brain to see this as happiness. The best example of this is scrolling on your smartphone. If this truly made you fulfilled, then the moment you stop scrolling you would be the nicest and most positive person in the world.

Man (Logic)

We know humans are conscious beings made in the image of God. People who try to be happy by imitating God in their own strength are people who look for opportunities to achieve something as an effect of stepping into tension and using their intellect. These people would answer the question above with: "get promoted, get a raise, win an award, get a degree, solve a problem," and basically "take on a challenge."

We all have this thought process in areas of our lives where we want to grow, and this means we have to push through the tension. We could see this level as profitable; however, we know that everything we physically create will eventually burn. While we can create something of value in the moment, we aren't producing anything of eternal value. When we begin to believe *we* are able to produce anything of eternal value, we could be moving toward a thought process that would prevent our salvation.

This thought process is the trickiest because it logically seems right when we can feel good about ourselves in the moment and even see others in this thought process as our example. However, God made our logical brain to get used to the level of the challenge.

Think about it, do you get exactly the same level of "happiness" when you accomplish the exact same achievement the second time? No. You have to achieve something greater to get the same happiness as you got when you achieved something less! This means that eventually, you will hit a level when you can't achieve something greater and then you will move to the animal thought process and may even move to the evil thought process if you live long enough. Ultimately, the logical, human thought process is deceptive because it looks good in the moment, but is unsustainable in the long term.

Remember, the equation for happiness is: experiencing effects that exceed your expectations. When you have experienced a level of happiness, your expectations rise, and the only way to experience happiness the next time is to exceed the previous level.

Notice, the thought processes line up with the approaches to problems involving people, which means they line up with the gold through iron levels on the wall. Absolve is evil and allows destruction to continue. Resolve is not unprofitable by treating the effects. This human thought process is the solve approach, and it can look good in the moment because it is logical; however, it will always lead to a worse condition in the long term. It looks like there is only one thought process that is not only eternal but also healthy!

Godly (Mada)

We saw that mada is the godly thought process. Mada is eternal because it dissolves problems. We have this thought process when we are completely confident in who we are, which means we can allow the Holy Spirit to flow through us (grace). Think back to the equation for happiness: experiencing effects that exceed your expectations. The only way to achieve long-term healthy happiness is to give up control (experience effects) and give up expectations of perfection (exceed expectations). Isn't this how we have described growing in grace and faith?

People who answer the question with: "respond to everything by being more who God created me to be" are in the mada thought process and have embraced their role in God's eternal plan. So, what is our conflict?

Conflict: You have a plan for your life that you logically believe will make you happy and the reality is it won't.

Resolution: Either continue to do your plan and end up in the lake of fire or admit your plan doesn't work and that you don't want to do your plan anymore. Instead, you want to give up control and allow God to do His plan through you by grace. This is confession and repentance, and immediately results in salvation, while continuing to grow in this leads to sanctification and reward.

We know every human that exists will either end up in the new Jerusalem with a godly thought process or the lake of fire with a destructive thought process. This means it is impossible to sustain a human or animal thought process indefinitely. You are either moving towards mada or evil. Now we can look at God's resolution.

God's Resolution

God brings about the resolution to His story through two ways. First, He operates through people who have given their will up to Him. These people are allowing more of God into the world (salvation) because God can't initiate bringing more of Himself into this world, and these people have a share in what God is accomplishing (reward). People can do this by prayer (giving their value to God), love, forgiveness, sharing the truth, and handling persecution well. The Bible says these people are born of God and are growing towards the mada thought process.

The second way concerns everyone else. God is giving to these people and not expecting anything in return (love), so when He forgives, justice credits God with the spiritual value (reward), while these people end up in the lake of fire. God can initiate bringing more of Himself into our world in response to justice crediting God with reward from these people not giving back to God (which would have been loving God). While these people are following one of the three thought processes other than mada, they will ultimately end up in the evil thought process. This leads to the following revelatory conclusion:

God is bringing about His will through everyone!

The reality is whether you are a believer or not, you are facilitating God's will. The only issue is whether you are working with God and will receive salvation and reward or you are working against God and receive the lake of fire. Notice, people who say they are against God still facilitate His will even though they are doing the opposite of His will.

I see it this way: God needs to accumulate a specific amount of spiritual value. God is accumulating spiritual value $100 bills through those who immediately allow God to do His will through them, while God is accumulating spiritual value pennies through those who are resisting God's will.

For instance, people who resist God's will by inaction are under mercy until they make a decision. When God increases the tension and they eventually choose to be outside the will of God, God gains value but it is relatively pennies compared to the person who immediately chooses to oppose God's will. God can immediately get His value in response to their decision.

Ultimately, the person who chooses to do God's will is immediately generating a tremendous value depending on whether they are in God's perfect, pleasing, or good will. Also, the value can accumulate based on everything that is accomplished in response to the person following God's will. This value is like a $100 bill compared to pennies from the person who demonstrates inaction.

Either way, God is accumulating value, so He knows (and we do too) that God's will is going to happen. The issue is how long it's going to take for God to accumulate the required spiritual value and that is completely dependent on people, not God.

If everyone had followed God's plan for their lives instead of their own plan, everyone would have ended up in the new Jerusalem, and no one would have ended up in the lake of fire. The required spiritual value would have been generated very quickly.

On the other hand, the number of people who hinder God's will determines how long it will take for God to obtain the spiritual value He needs to bring about His will and how many people are going to spend eternity in the lake of fire. This is our fault, not God's.

This is how God can know He will accomplish His plan according to His will well enough to give prophecy.

Summary

To summarize God's story, we see it's really about how Jehovah has been and continues to be rejected by the very people He's trying to help, but eventually ends up spending eternity intimately hanging out with a select group of Righteous people.

Paul and Jesus gave us examples of what it looks like to share in Jehovah's story:

- They both lived according to the influence from the Father via the Holy Spirit (mada).
- They both experienced active opposition to their respective missions, missions that were Righteous and facilitated God's goal!
- They were persecuted and rejected by religious authorities, the people everybody thought represented God.
- They were persecuted because the groups of people they spoke to felt uncomfortable because the information was different, which the groups of people verbalized as wrong and bad!
- They were persecuted because they were good, because they were Righteous!

How would you describe the journey you are on right now?

Dr. Joel Swokowski's Commentary

Tips for Applying this Topic

Write out what **you** believe God's story is…and share it with **Him**! Allow me to be more specific. We are talking to God the Father, and His name is Jehovah. I have found that referring to Him by name helps facilitate the fact that you are speaking with a specific Person.

I know, Jehovah already knows His story, but the point isn't to teach Jehovah, it's to share with Him. When you tell His story to Him, He will learn about you from the things in His story that are important to you! I highly recommend that once you have it written, the sharing part is done out loud. I remember the first time I did this. I was struck by how emotional of an experience it was!

If you need some help outlining Jehovah's story, check out: Daniel 9, Ezra 9, Nehemiah 9, and Acts 7 (Stephen). These are all instances where someone told Jehovah's story. Read them and see if you can find Jehovah's story within those chapters. Take notes to find the characters, conflict, and resolution. Take notes on which part of Jehovah's story you connect with the most, and which is the most inspiring (or convicting!).

Investing in God's Plan

We all have an opportunity now to invest into the greatest economy that will ever exist: God's plan. The way to have a share in bringing about God's plan is through participating in His will. There are three contextless principles that, when applied, always facilitate God's will. Those principles are:

1. Grace: the divine influence upon the heart, and its reflection in the life (mada).
2. Love: giving a value without expecting anything in return from the person to whom you gave.
3. Leadership: facilitating the purpose and progress of others.

Contextless means that each can be used in every situation by anyone regardless of their uniqueness, and when they are utilized, they make every situation generative. To be contrastive, I would simply ask:

1. What situation would be worse if you lived by grace?
2. What situation would be worse if you were loving others?
3. What situation would be worse if you were leading others?

Simple to see, the answer to those contrastive questions is a resounding "None."

Another benefit to living these contextless principles is they lead to your reward and/or sanctification. I recently had an interaction with my sister-in-law Heather concerning this very topic. She shared the following and asked me for clarity:

"So, we learned that God's Righteousness through us leads to salvation, and our righteousness leads to reward. The confusion that I had was around how reward all works when we allow God to work through us. Jesus tells us to store our reward up in heaven but we also want to be taking direction from God."

Here was my answer:
"It all depends on the origin of the behavior: was it God telling you to do something? OR, Was it you bringing something to God?

For example:
God says to you: Have a conversation with your daughter about sex.

This is salvation. God is the origin of this. He is telling you to do something and you allowing Him to work this influence through you would result in you being obedient and would be on the salvation side. Depending on the result of this conversation and what is created out of it, there could be reward accrued. However, the specific conversation that God told you to have is according to the salvation model.

OR

You say to God: I want to have a conversation with my daughter about sex…would You help me?

This is reward. You were the origin of this. You want to talk to her, but you are still including God in this conversation. This would result in reward and not impact your salvation. The complexity of the reward model, though, is that you don't have to include God in the conversation with her. It would still impact your reward and not your salvation…but asking God to work through you in this conversation would result in your daughter benefiting more and you having more reward…because God is better than you at bringing a value to people!

It ALL comes down to who was the origin of whatever you're going towards: you or God?"

The contextless principles will lead to reward if you initiate and sanctification if it's grace (God initiates). Either way is a benefit to you!

The Ultimate Approach to God's Will

The ultimate approach to God's will is rest. Resting is ceasing from the occupation of being a first cause. God did this on the seventh day, and now we are encouraged to enter into His rest. If you wanted to sum up and/or generalize the contextless principles of grace, love, and leadership, you could with the doctrine of rest. The point of rest is to give up control and put God back in the position of being a first cause through you.

Resting is critical for God's will to happen. The more people rest, the quicker God's will happens. We are responsible for the efficiency of God's will.

Afterword

THE FIRST OF these two books showed how God brings about His eternal plan, which we are calling God's will. The long answer to God's will would involve reading and interpreting the Book of Job because not only did Job explain God's Nature, he explained how God brings about His will. Job is accumulating spiritual value for presenting the answer and at the expense of people misinterpreting, ignoring, or being unbelievers with respect to his book.

The short answer is God brings about His will through people. In fact, God is bringing about His will through everyone because everyone either has spiritual value they are allowing God to use to bring about His plan through prayer or is allowing God to acquire spiritual value to bring about His plan at their expense.

The first group are believers: people who have given their will up to Him. These people are allowing more of God into the world through grace (salvation) and these people have a share in what God is accomplishing (reward).

The second group are unbelievers: people God gave to and didn't expect anything in return, so when He forgives them for not giving back, justice credits God with the spiritual value (reward), while these people end up in the lake of fire.

In both cases, God is presenting the truth in love to people. Believers love God by allowing the truth (mada) to work through them to accomplish His will, resulting in spiritual value, which the believer has a share in. Unbelievers don't love God, and God forgiving these people to whom He gave the truth in love, results in God acquiring spiritual value through justice.

Either way, God is able to achieve enough spiritual value to bring about His plan, it's just that the unbeliever route takes more time and results in more people ending up in the lake of fire. It would seem that everyone would want to know which type of person they really are.

We've also moved far enough away from the wall to see there is an image of large denominations of spiritual value in the gold level, while there is an image of small coins in the iron level. Also in the iron level is an image of the lake of fire along with the phrase "Your Plan." Finally, each level has the corresponding thought process: mada (gold), human (silver), flesh (brass), and blood (iron).

Book 2 will continue to fill in the rest of the wall by showing how we can determine if we are in God's will to the point we can help God bring about His eternal plan, what we have been calling "the party." It turns out, there are some crucial areas where people don't realize they are opposing God's will! This will also get uncomfortable because we are filling in a system and some things may not make sense until you have all the parts.

We have said the party itself is the ultimate in intimacy and communication by exchanging with others through our AREs. Let's begin by looking at intimacy and communication.

Dr. Joel Swokowski's Commentary

Humans are born dependent. A baby will not survive without the care of his parents. Perhaps there is an image of a baby in the brass level to represent this. As it relates to God, when a person admits their spiritual dependence on God, this is representative of salvation.

At some point, we become anxious to demonstrate we are no longer a child. This is independence. Perhaps there is an image of this in the silver level. We have covered how teenagers are so anxious to prove they are their own person, they tend not to take advice from others. Likewise, when we become spiritually restored and are able to bear our own burdens, we can begin to grow in our uniqueness and holiness. This is sanctification.

Hopefully, the teenager will realize that maturity is not independence because no one can make it through life on their own. Adulthood is when we realize independence is not generative, and the only way to achieve generativity is through interdependence, which is only relying on people who rely on us, like a council meeting. Perhaps there is an image of this in the gold level.

This is the stage this book is trying to help you reach by stating the uncomfortable conclusion: God relies on us to bring about His will, and if we allow God to facilitate His will through our uniqueness, it will result in our healthy happiness, which also involves being interdependent with other people by exchanging spiritual value through our uniquenesses.

We need to stop hindering God's will by doing our own plan for happiness. Doing this is a catalyst for controlling others which makes us feel less in power over time, which causes us to try to control even more. In Book

2, we are going to see that if I'm questioning whether I'm still doing my own plan for happiness, I can ask myself, "Did I try to make it happen?" If so, then it wasn't God, even when I get what I think will make me happy.

When we start bringing value to other people and accepting value from others, we are interdependent. This is representative of repair, discipleship, and leadership. This person's uniqueness is focused on helping the mission of the greater system.

God's will is interdependent in nature. We need God to be part of His eternal plan. God needs us for His plan to come to fruition.

The Party

Chapter 1 reminded us of the following:

"The analogy guiding each of these three editions was presented in the following manner at the beginning of Modeling God, Chapter 1: The Invitation.

Imagine you are invited to a party where you will be able to participate in any sensual pleasure you desire for as long as you want."

Modeling God also taught about the Physical ARE. Another way to understand the Physical ARE: it is your plan for happiness.

Think of the invitation like this: you have been invited to a party (God's party), but in reality, you are already headed towards a party. The party you are going towards is aligned with your plan for happiness. This book helps you understand God's party so you can intentionally choose the party you want to attend, for eternity!

Book 2

Determining God's Wills

PART ONE

Intimacy and Communication

INSIDE

Chapter 1: Intimacy 231

Chapter 2: Covenants 248

Chapter 3: Communication 260

Chapter 4: Repair 273

Chapter 5: Transformation 286

CHAPTER 1

Intimacy

WE HAVE BEEN calling God's eternal plan "the party." We have said the party itself is the ultimate in intimacy and communication by exchanging with others through our AREs. If we want to be able to determine if we are in God's will to the point we can help God bring about His eternal plan with profitability through justice, we need to look more closely at intimacy and communication. This will show us if we are helping God's will or hindering it with our traditions. Let's begin with intimacy.

The way we measure intimacy is according to sharing. Sharing is an exchange that is mutually beneficial. Let's look at the first four levels of sharing.

> 1. **Interaction — superficial sharing in one area.**
> Example: A stranger telling us that a drink we have is going to fall off a table if we don't move it. This information is limited to one area and usually consists of a *what*, which means anyone can participate at this level.
>
> 2. **Connection — deep sharing in one area.**
> Example: Talking about work with a coworker. Another example is talking about a favorite movie, TV show, or musician with

another fan. While this information is limited to one area, it will go past the *what* to include the *why* and *how*. The discussion will be deeper than the *what*. Notice, two coworkers could talk about the simplest fact (*what*) about their company, and if I don't work at that same company, I may not know what they were talking about.

3. Relationship — deep sharing in at least one area and superficial sharing in other areas.

Example — Relatives and people you grew up or went to school with are good examples. Notice, this is deep sharing in one area, but the sharing isn't limited to that one area. For example, if someone at work (connection) asked you what you are going to do this weekend, they are assuming you are at relationship with them because they expect you to share information about yourself other than something related to work.

The word relationship was not in the Bible. In fact, the Greek word for relationship gets translated as ratio, reference, and (sometimes) model in English. Think about it, the word relationship implies the level to which you can give and take. For example, "Where is that person in relationship to you?" implies both of the people involved are static. The implication is that neither person is growing. Then what about people asking about your personal relationship with Jesus?

> *That which we have seen and heard declare we unto you, that ye also may have fellowship with us: and truly our fellowship is with the Father, and with his Son Jesus Christ.* (I John 1:3)

We aren't supposed to have a relationship with the Father and Jesus, we are supposed to have fellowship. The objective of fellowship is to be deeply known and to deeply know another. It is all about giving and growing in giving!

4. Fellowship — deep sharing in all areas.
This means fellowship is complete sharing with others. That is, when a person is willing to share all his information with another person and also tells that person they can initiate questions about his life.

The process of getting to fellowship involves progressing through other levels of sharing. In fact, the easiest place to see this is with romantic comedy movies. Every rom-com movie can be distinguished by using these four levels of intimacy.

Interaction occurs when the couple meets. It is the moment each becomes aware of the other, usually through an introduction and sharing of names. Connection is when the couple finds a shared interest they are both passionate about. Relationship is the montage portion of the movie: walking, going to dinner, riding bicycles, etc. Fellowship is when the couple commits to being totally open with each other and usually occurs in response to a conflict.

Putting this all together, we could summarize a film by saying: They met on the bus and found out they both loved Mozart. They spent weeks together until she thought he was seeing someone else. When she confronted him, she found out it was his sister who he talked into giving her the big career break she has been working towards.

Movement In Sharing Levels

Profitability ought to guide us as to the level we are at with other individuals. Since this goal of profitability is intrinsic to everyone, people move through the levels of sharing in a fluid manner. As people experience profitable exchanges with others, they will look for the opportunity to move to a deeper level of sharing. However, a person can go back to a previous level of sharing if their intimacy with the other person is unprofitable.

There are three guidelines to ensure profitability:

1. In order to progress further, trust has to be built. (Trust is built the same way faith is built: understanding and experience.)
2. The appropriate level of sharing is determined by profitability.
3. If people aren't profitable at a level, they ought to go to the next level of sharing that is more superficial and recheck profitability.

Each time there is growth and movement in sharing levels, it is an effect of making the other person more uncomfortable by requiring both people to share more of themselves. If these uncomfortable opportunities are handled well, then trust is increased. If these uncomfortable opportunities are not handled well, then trust decreases.

For example, if someone uses something you shared with him against you, you will trust him less, share less, and the level of sharing moves to a more superficial level. It then takes more than one well-handled experience to overcome the damage from the one poorly handled experience.

The stress in sharing can always be attributed to people disagreeing on the appropriate level and/or direction of sharing. Direction means one of the individuals wants to move to a different level of sharing from what the other person wants.

You can only remain profitable at a specific level of sharing if both people are profitable at that level. For example, if they make it unprofitable for another person to remain at relationship, then they can't demand the other person maintain sharing at the relationship level.

Some would say this is making life more complicated. While this model does not create these levels in each interpersonal exchange, these levels of intimacy and this process occurs whether you are aware of it or not. This model just helps you understand where you are at, why there is or isn't a problem, and where you are going to end up.

Remember, sharing is an exchange (right *what*) that is mutually beneficial (right *how/why*). We have said stress occurs due to disagreements in the levels of sharing. Most of the time, this involves the relationship and fellowship levels. Let's look more closely at the difference between relationship and fellowship.

Relationship vs. Fellowship

Here is a short list of relationship characteristics:

- Relationship allows us to initiate superficial sharing about the other person.
- Relationship requires us to accept the response or lack of response we get from the other person.
- Relationship does not allow us to initiate deep sharing about the other person. It only allows us to initiate deep sharing on ourselves.

The objective of fellowship is to be deeply known and to deeply know another. There is a willingness between the two people to share deeply. Here is a short list of fellowship characteristics:

- Fellowship allows us to initiate deep sharing about the other person. The reason why we are initiating deep sharing about another person is for their benefit (growth) and to deeply know the other person. In fact, fellowship would cause us to initiate deep sharing on ourselves in hopes of being deeply known.
- Fellowship is love: giving a value and not expecting anything in return.
- Fellowship requires the other person to give an open, truthful, and vulnerable response.

NOTICE: Honest means factual; only a right *what*. Truthful is a right *what* with a right *how/why*.

NOTICE: In fellowship, the individual commits to sharing everything and invites the other person to initiate personal discussion.

NOTICE: The goal of all of this is growth (profitability). The only way to get better is to actively look for areas of improvement and change them. Fellowship results in growth (profitability) for everyone!

Example #1: Parent and child

- Parents want to operate at fellowship in their approach to the child. The parents want to question the child and demand an honest answer.
- Parents don't want to personally give the child access at fellowship. The parents don't want the child to question them or have to give the child an honest answer.

Result

- The child grows up with a skewed perspective of right and just (God) if the parents continue this way. The child thinks the ultimate authority (God) operates this way.
- The child grows up without a perspective of right and just (God) if the parent goes to relationship because the goal isn't fellowship.

Example #2: God and us

- God wants to be deeply known by others.
- God wants to operate at fellowship in His approach to us.
- God wants to question us and get a truthful answer.
- God wants to personally give us access at fellowship.
- God wants us to question Him and He wants to give us a truthful answer.

The conflict and lack of profitability come when:

- we don't want God to question us.
- we don't want to question God.
- we don't want a truthful answer from God.
- we don't want to give God a truthful answer.

Today Christians usually speak of their personal relationship with Jesus. Have we replaced God's goal of fellowship with man's goal of relationship? Are we teaching that God considers the goal to be a relationship? How does this affect us with other believers?

> *That which was from the beginning, which we have heard, which we have seen with our eyes, which we have looked upon, and our hands have handled, of the Word of life; (For the life was manifested, and we have seen it, and bear witness, and shew unto you that eternal life, which was with the Father, and was manifested unto us;) That which we have seen and heard declare we unto you, that ye also may have fellowship with us: and truly our fellowship is with the Father, and with his Son Jesus Christ.* (I John 1:1-3)

This passage said the reason they shared Christ with others was so they could have fellowship with them! Their goal was fellowship.

This Bible verse shows you cannot reach fellowship with another person until you have both reached fellowship with God. You cannot trust another person to look critically at you until they have recognized they are not perfect and actively invite God to look critically at them. It may look like fellowship, but it will end up being destructive.

Also, people who attempt to prevent believers from having fellowship with other believers, for any reason, are hindering the word and will of God. They are hindering profitability through man-driven control.

In fact, what do the levels of sharing between people and God look like?

Sharing Levels: people to God

- **Interaction** — aware of God
- **Connection** — praying to God in time of crisis
- **Relationship** — born again; begin the process of opening up in all areas to God
- **Fellowship** — strong commitment to growth in taking direction from God in all areas

So, it looks as if the reason we don't have fellowship in the church is because we don't first have fellowship with God and then with each other.

You can have a relationship without love. You cannot have fellowship without love. The Bible says that the only reason we are able to love is because we choose to let God work through us. This is why we can't have fellowship with each other until we first have fellowship with God. We have to do Jesus' first commandment (love God) and the effect is Jesus' second commandment (love our neighbor).

How do we get to fellowship with God?

Fellowship With God

In the previous section, we saw that we cannot have fellowship with others unless we first have fellowship with God. The other requirement for achieving fellowship is to know your ARE; who God created you to be.

Remember, sharing is a right *what* (exchanging information) with a right *how/why* (for the benefit of another/for understanding). The objective of fellowship: To be deeply known and to deeply know another.

You will be limited in how much you can let others know you by how much you know yourself. You also need to know who God is. We saw that God's Nature is Right-Right and Just, so it isn't enough to understand *what* God does. You also need to understand *how/why* God does *what* He does.

Getting to fellowship with God is a process. Whether we are conscious of it or not, God wants to be in fellowship with everyone. God is constantly working with everyone in an attempt to get to fellowship. This is right and just. The only reason a person isn't at fellowship with God is the person consistently chooses not to be in fellowship with God.

Fellowship with God is a three-step process. I have seen this process play out consistently both in biographies and personally talking with people. Several times when people have told me they have taken the first step, I have warned them to watch out for the second step, only to have their eyes grow wide and tell me that it is exactly what they are going through now!

This process of going to fellowship with God is painful to our flesh. Some people call it "The Breaking Process." However, I have seen that when people understand the process, they are able to intentionally work through it and feel like they are not alone or insane.

We saw that relationship with God begins when a person is born again. Jesus' first commandment says to love God with all your mind, heart, soul, and strength. This is fellowship. The word all means completely sharing yourself with God in everything. This will require God to break the person from controlling or isolating certain areas of their life.

When God breaks a person, the goal is to help the person become who God made them to BE. This is done for the benefit of the person. This is love.

The outside of the person is the DO (what you do) and the HAVE (what you have). The BE (who God made you to BE) is on the inside. The flesh has to be broken in order for this BE to come out. I called this the ARE (who you ARE; who God created you to BE).

If we are soft and young, the breaking process is easier. If we are constantly staying broken, the process is easier. If we are hard and/or have spent years being unbroken, the process is painful. *Modeling God* showed us how to begin to find out our ARE in general terms. We ought to continue to find out our ARE in more specific terms. God will not show a person exactly who they are made to BE until they completely trust God. This proof of trust is a process.

The first step in the process is God pushing the person to get him to admit he is wrong. This (and every) step is actually a decision. God is essentially saying, "Do you admit you are wrong?" This is what Joel called a full confession in the commentary for *Modeling God*.

The person can essentially say, "no" with his actions, or he can avoid the question through drugs, staying busy, blaming others, etc. This is pride. People can get so devastated by this they turn completely away from (reject) God, but it is the person's fault. God stays on the person and ramps up the stress through justice until the individual finally says, "Yes, I am wrong."

The second decision that God is pushing the person towards, for their benefit, is for the individual to admit he needs to rely completely on God's help. God is essentially saying, "Do you admit you need to rely on Me completely?" Doing the action (relying on God completely) is what Joel called a full repentance in the commentary for *Modeling God*.

Again, God stays on the person and ramps up the stress through justice until the individual finally says, "Yes." Likewise, people can get so devastated by the pain that they turn away from (reject) God.

Notice, the first two questions are really a deepening of the relationship. Full confession and full repentance is the most complete way to be born again. The next decision is the most brutal. Remember, God is trying to find out if the person trusts Him completely. God asks the person if they will follow Him regardless of whether God takes everything.

This can be presented by God as a series of questions:

- Will you follow Me if I take away your job?
- Will you follow Me if I take away your money?
- Will you follow Me if I take away your friends?
- Will you follow Me if I take away your kids?
- Will you follow Me if I take away your spouse?

Look at it this way: You occupy a specific space. Everything else in your life occupies a space around you. God's goal is to rebuild you into something else, and you will most likely occupy a new space. That will create stress between you and everything else in your life. This will look to you like God is taking these things away because He is causing the stress. Actually, it is everything around you that is choosing to adjust or leave you.

God wants to know, before He begins the process of making you into exactly who you were made to BE, if you are completely trusting Him, that is, if you want to go to fellowship with Him. Jesus said that only fools build a tower or go to war without first figuring out if they are able to complete the task (Luke 14:28). God is not a fool. He is not moving on His part until the individual finishes his part (three steps) by stating his will that he will still follow God no matter what.

I have told people to take their time with this step. They need to imagine what it would look like for each area to reject them. This is not something that ought to be rushed. This is a brutal process.

When the person can say "yes" to each question, then they are completely broken, and God can access their ARE and begin the process of building them up into what God wants them to BE. They can have fellowship with God, and they can have fellowship with others who have fellowship with God.

Unfortunately, some people must lose everything before they choose to have fellowship with God. There are plenty of books that say we are broken people and that we need to be broken. What is the process these books describe? Can you do their process intentionally?

Ultimately, everything comes down to whether we are doing God's will according to His word resulting in profitability, or we are doing man-made tradition and deceiving ourselves that we are in God's will.

Can you see why we don't have fellowship in church? Can you see why we don't have fellowship in marriages? It starts with each person having fellowship with God. Are you helping or hindering God's will with respect to intimacy?

I believe it is pretty easy to see each of these levels on the wall: fellowship (gold), relationship (silver), connection (brass), and interaction (iron).

As intense as the fellowship level is, there is a fifth level of intimacy…

Dr. Joel Swokowski's Commentary

Council Meetings Implication

If facilitating God's will includes us embracing council meetings, then God's will being accomplished requires me to be in a position to have other people speak into my life. What is a council meeting if not a group of people I have asked to show me perspectives and give me ideas that I'm not seeing myself?

This chapter gave us clarity on a previous chapter. The leader of a council ought to be at least at the relationship level with their council members. The leader will be sharing with the council, and that leader needs to, at the very least, value them for their counsel in a specific area, the area where the leader is willing to share deeply.

A council at interaction and connection likely wouldn't know the leader well enough to give the counsel he/she needs, nor would they be in a position where they'd be willing to speak into their life. A council at fellowship would create an environment with even greater wisdom than relationship since the members would also be hearing from God for the leader. Relationship would be the minimum for a profitable council, while fellowship would be ideal. Let's look a bit closer at fellowship.

Fellowship

Fellowship was the Greek word "koinonia" G2842.
> koinonia — from 2844; partnership, i.e. (lit.) participation, or (social) intercourse, or (pecuniary) benefaction.

Social Intercourse: that's a conjunctive!
> Freedom: Intercourse
> Limitation: Social

- Intercourse can be seen as an interaction between two or more people, a quantitative aspect of fellowship.
- Social can be seen as the quality of interaction, the depth of mental, emotional, spiritual, and even physical (with one exception).

Social intercourse (fellowship) is the deepest level of intimacy possible, at the social level.

This definition of fellowship may bring a higher standard to this principle than you had previously known. I've even seen people be taught this and realize they aren't in fellowship with anyone. The reality is, learning this definition didn't stop them from being in fellowship with anyone, they just weren't aware of the true standard for fellowship.

Now that they have the right definition, they have two choices:

1. Use that new information and standard of fellowship to grow their intimacy with God and others.
2. Keep their contradictory definition of fellowship to rationalize the lack of intimacy in their life.

Confession and Repentance

Confession is admitting what we did that was wrong, the abuse we caused. In the commentary for *Modeling God*, I wrote that a full confession involves three parts:

1. I know what I did was wrong. "What I did" is the abuse, and it needs to be stated.
2. I know why I did it. The reason why the abuse occurred also needs to be stated.
3. I don't want to do it again.

We saw in *Modeling God* that repentance begins by asking what needs to be given or done to make up for the wrong, and then doing it. The goal is that repentance ought to make the person happy the abuse occurred.

Example: If I slap you, you won't be happy. However, if I give you a million dollars because I slapped you, every time you tell people about me slapping you, you will do it with a smile.

Now clearly, it would have been best if the abuse never occurred, but once it does, repentance is the path to making the situation better.

A full repentance would not only involve the person repenting, but also going to everyone that is aware of the abuse and letting them know how excellent the person who got abused actually is. Notice, people may not have thought much of the person who got abused, but now, as an effect of being abused, people think more of them. The abused person truly ends up better because of the abuse.

The Greatest Commandment

Jesus taught that the greatest of the commandments was actually in two parts:

> *And one of the scribes came, and having heard them reasoning together, and perceiving that he had answered them well, asked him, Which is the first commandment of all?*
>
> *And Jesus answered him, The first of all the commandments is, Hear, O Israel; The Lord our God is one Lord: And thou shalt love the Lord thy God with all thy heart, and with all thy soul, and with all thy mind, and with all thy strength: this is the first commandment.*
>
> *And the second is like, namely this, Thou shalt love thy neighbour as thyself. There is none other commandment greater than these.* (Mark 12:28-31)

We saw in *Modeling God* that Jesus actually came to earth with two messages: how to get saved and how to gain reward. It's incredible to see both messages reflected in one answer to a question Jesus was asked. Actually, it seems like He couldn't even answer the question of what is the *first* commandment of all without keeping both the salvation and reward messages connected, yet separate.

See, the simplest form of teaching a person about how salvation and reward are connected is through this teaching by Jesus:

1. We first love God as a cause (mada, grace, salvation),
2. Then God flows through us to love others as an effect (reward).

So, even when we start emphasizing how we're meant to grow our intimacy and communication by exchanging with others through our AREs, it still all comes down to first loving God. This is most evident in the fellowship level, and the one coming next.

CHAPTER 2

Covenants

ACCORDING TO THE Bible, covenants are the fifth and most intimate level of sharing. In *Modeling God* we saw that a covenant is when two or more people give each other complete access to all they HAVE, DO, and ARE.

In the Old Testament, the word for covenant is "beriyth" (H1285). The entry in Strong's Concordance says:

from #1262 (in the sense of cutting [like #1254]); a compact (because made by passing between pieces of flesh): - confederacy, covenant, league.

When looking at the Bible, there seems to be five parts to a covenant, and they don't have to occur in a particular order. In rough terms, they are:

1. go through ("passing between") something
2. bloodshed
3. sharing
4. agreement
5. death

This sounds violent, but remember, the root for the word covenant comes from cutting. Also, covenants require progress. The profitability of the covenant can be increased by driving all five parts.

Notice, these five parts should be done by us with God: being in agreement with (accepting) God, making the spiritual pre-eminent, sharing, going through tough times, and giving God everything (spiritual and physical).

There is a difference between not driving a part and violating a part. Violating a part of the covenant breaks the covenant (which we will cover later). Not driving a part is not achieving everything possible in that part and leads to less profitability. Let's look at the covenants mentioned in the Bible.

Noah

The first covenant mentioned by name in the Bible was between Noah and God. It was different from all the other covenants that were mentioned in the Bible.

> But with thee will I establish my covenant; and thou shalt come into the ark, thou, and thy sons, and thy wife, and thy sons' wives with thee. (Genesis 6:18)

First of all, this covenant occurred after Noah completed his part:

1. **go through something:** Noah went through the Flood
2. **bloodshed:** sacrificed animals
3. **sharing:** gave years of his life doing God's will
4. **agreement:** agreed with God's new plan
5. **death:** everyone on earth except those in the ark

The rest of the covenant was completely dependent on God and is still ongoing, which explains why Noah was mentioned as one of three people who had tremendous spiritual value in Ezekiel 14. God gave us a token (rainbow) to remind Him to keep His promise never to flood the earth again.

Abram

The second covenant mentioned by name was with Abram and God.

> *And he said unto him, Take me an heifer of three years old, and a she goat of three years old, and a ram of three years old, and a turtledove, and a young pigeon. And he took unto him all these, and divided them in the midst, and laid each piece one against another: but the birds divided he not. And when the fowls came down upon the carcasses, Abram drove them away. And when the sun was going down, a deep sleep fell upon Abram; and, lo, an horror of great darkness fell upon him. And he said unto Abram, Know of a surety that thy seed shall be a stranger in a land that is not theirs, and shall serve them; and they shall afflict them four hundred years; And also that nation, whom they shall serve, will I judge: and afterward shall they come out with great substance. And thou shalt go to thy fathers in peace; thou shalt be buried in a good old age. But in the fourth generation they shall come hither again: for the iniquity of the Amorites is not yet full. And it came to pass, that, when the sun went down, and it was dark, behold a smoking furnace, and a burning lamp that passed between those pieces. In the same day the Lord made a covenant with Abram,* (Genesis 15:9-18a)

Here are the parts of Abram's covenant:

1. **go through something:** Abram had left his land to follow God
2. **bloodshed:** sacrificed animals
3. **sharing:** gave years of his life doing God's will
4. **agreement:** agreed with God's plan
5. **death:** sacrificed animals

This act was very intentional. Not only were all five parts of the covenant covered, but look at the imagery! Abram put bloody pieces of heifer, goat, and ram across from each other. Then there was a smoking furnace and a burning lamp that passed between those fur-covered bloody pieces of flesh. (This aspect of a covenant was covered in the Strong's definition.) The covenant was consummated.

Moses

Which things are an allegory: for these are the two covenants; the one from the mount Sinai, which gendereth to bondage, which is Agar. (Galatians 4:24)

Let's look at these two covenants Paul mentioned in more detail. The covenant Paul wrote about originating from Mt. Sinai was the covenant God cut with Israel through Moses. In Exodus 19, Israel arrived at Mt. Sinai three months after leaving Egypt. God told Moses to tell Israel:

Now therefore, if ye will obey my voice indeed, and keep my covenant, then ye shall be a peculiar treasure unto me above all people; for all the earth is mine: (Exodus 19:5)

Covenant List:

1. **Go through something** — The Israelites had gone through the Red Sea on dry ground. They passed between a wall of water on each side. *And the children of Israel went into the midst of the sea upon the dry ground: and the waters were a wall unto them on their right hand, and on their left.* (Exodus 14:22).
2. **Agreement** — *And all the people answered together, and said, All that the Lord hath spoken we will do. And Moses returned the words of the people unto the Lord.* (Exodus 19:8)
3. **Sharing** — Chapters 20 through 23 of Exodus recorded God sharing His Law and promises of blessings with the people.
4. **Death** — *And he sent young men of the children of Israel, which offered burnt offerings, and sacrificed peace offerings of oxen unto the Lord.* (Exodus 24:5)
5. **Bloodshed** — *And Moses took half of the blood, and put it in basons; and half of the blood he sprinkled on the altar. And he took the book of the covenant, and read in the audience of the people: and they said, All that the Lord hath said will we do, and be obedient. And Moses took the blood and sprinkled it on the people, and said, Behold the blood of the covenant, which the Lord hath made with you concerning all these words.* (Exodus 24:6-8)

Jesus

Paul also wrote about a covenant with Jesus. The Greek word for covenant used in the four Gospel accounts was translated as testament.

From Strong's Concordance:
#1242 diatheke, from #1301; prop. a disposition, i.e. (spec.) a contract (espec. a devisory will): - covenant, testament.

For instance, when Jesus held up the wine at the Last Supper, Matthew 26:28 stated:

For this is my blood of the new testament, which is shed for many for the remission of sins.

Covenant List

1. **Go through something** — The events that led up to the crucifixion. *As many were astonied at thee; his visage was so marred more than any man, and his form more than the sons of men:* (Isaiah 52:14)
2. **Sharing** — We share in Jesus' death (Romans 6:4-5) and Jesus shared the Holy Spirit with us.
3. **Agreement** — Jesus willingly endured the cross. *Therefore doth my Father love me, because I lay down my life, that I might take it again. No man taketh it from me, but I lay it down of myself. I have power to lay it down, and I have power to take it again. This commandment have I received of my Father.* (John 10:17-18)
4. **Bloodshed** — Jesus shed His blood on the cross and as we saw above (Matthew 26:28), He identified the blood of this covenant.
5. **Death** — Jesus' death on the cross.

Breaking vs. Ending a Covenant

As we saw, God cut a covenant with Israel through Moses. Israel continued to worship other gods, which broke the covenant with God. We saw that Solomon reestablished the covenant when he built the temple.

> *Thus Solomon finished the house of the Lord, and the king's house: and all that came into Solomon's heart to make in the house of the Lord, and in his own house, he prosperously effected. And the Lord appeared to Solomon by night, and said unto him, I have heard thy prayer, and have chosen this place to myself for an house of sacrifice. If I shut up heaven that there be no rain, or if I command the locusts to devour the land, or if I send pestilence among my people; If my people, which are called by my name, shall humble themselves, and pray, and seek my face, and turn from their wicked ways; then will I hear from heaven, and will forgive their sin, and will heal their land. Now mine eyes shall be open, and mine ears attent unto the prayer that is made in this place. For now have I chosen and sanctified this house, that my name may be there for ever: and mine eyes and mine heart shall be there perpetually. And as for thee, if thou wilt walk before me, as David thy father walked, and do according to all that I have commanded thee, and shalt observe my statutes and my judgments; Then will I stablish the throne of thy kingdom, according as I have covenanted with David thy father, saying, There shall not fail thee a man to be ruler in Israel.* (II Chronicles 7:11-18)

God even mentioned that He covenanted with Solomon's father David. However, God also explained what would happen if Israel didn't keep up their end of the covenant and broke it:

> *But if ye turn away, and forsake my statutes and my commandments, which I have set before you, and shall go and serve other gods, and worship them; Then will I pluck them up by the roots out of my land which I have given them; and this house, which I have sanctified for my name, will I cast out of my sight, and will make it to be a proverb and a byword among all nations. And*

> this house, which is high, shall be an astonishment to every one that passeth by it; so that he shall say, Why hath the Lord done thus unto this land, and unto this house? And it shall be answered, Because they forsook the Lord God of their fathers, which brought them forth out of the land of Egypt, and laid hold on other gods, and worshipped them, and served them: therefore hath he brought all this evil upon them. (II Chronicles 7:19-22)

We know that the kingdom split after Solomon died, and about 250 years later, Israel was taken into captivity. This occurred while Hezekiah was king of Judah. We also saw that even though Hezekiah's son Manasseh led Judah into idol worship and was the reason Judah would go into captivity, this didn't occur until after Manasseh's death and during Jeremiah's time. Let's look at how God responded to Israel very early during Jeremiah's time.

> Go and proclaim these words toward the north, and say, Return, thou backsliding Israel, saith the Lord; and I will not cause mine anger to fall upon you: for I am merciful, saith the Lord, and I will not keep anger for ever. Only acknowledge thine iniquity, that thou hast transgressed against the Lord thy God, and hast scattered thy ways to the strangers under every green tree, and ye have not obeyed my voice, saith the Lord. Turn, O backsliding children, saith the Lord; for I am married unto you: and I will take you one of a city, and two of a family, and I will bring you to Zion: (Jeremiah 3:12-14)

In this passage, married was the word "bâ'al, baw-al'; a primitive root; also as denomitive from H1167 to be master." God was saying He was the master of the covenant. Why? Because Israel had broken the covenant.

When one of the people in the covenant violates the terms of the covenant, the covenant is said to be broken. The person who didn't violate the covenant is now the master of the covenant. They have the choice to continue the covenant or end the covenant.

Israel continually broke the covenant from Solomon's time to Jeremiah's time, yet God chose to continue the covenant each time. In fact, the entire point of the Book of Hosea was that God compared Israel to a bride who chose to continue to act as a harlot, yet God would forgive her and stay married to her. Likewise, God would continue the covenant with Israel if Israel would stop worshiping idols. Israel chose to continue chasing other gods and it wasn't long after that they went into captivity.

We know how this turned out with Jeremiah because it was during Jeremiah's time that God said for Israel not to pray to Him because He wouldn't hear their prayers. Doesn't it sound like God ended the covenant? We will see in a later chapter why God did this.

Summary

We now see there is a fifth level of sharing: Covenant. This fifth level is consistent with the guidelines for profitability.

Remember:

- You can only remain profitable at a specific level of sharing if both people are profitable at that level.
- This model does not create these levels in each interpersonal exchange; these levels and this process occur whether you are aware of it or not.
- This model just helps you understand where you are at and where you can potentially go.

We now see how the levels of intimacy correspond to the individual and God:

- **Interaction** — aware of God
- **Connection** — praying to God in time of crisis
- **Relationship** — born again, begin the process of opening up in all areas to God
- **Fellowship** — strong commitment to growth in God in all areas
- **Covenant** — in heaven

We have covered intimacy and sharing. Let's look at communication.

Dr. Joel Swokowski's Commentary

Since the next chapter is about communication, I would like to set it up by sharing the following story.

My friend Morgan and I met in 2009. We were both part of a home church and ministry that has since expanded and multiplied, serving many people and regions. It's been a blessing getting to know her and working with her. When we met, we hit it off immediately. We became great friends and even took the intentional step of growing in fellowship with each other. It was amazing, and the interactions with Morgan are among the greatest to have taught me what it really means to be in fellowship with another person…and what fellowship is not.

We both knew that part of fellowship is giving each other permission to speak into each other's lives. And that is what I did. If she spoke a word wrong, I'd call her on it. If I saw a contradiction, I'd expose it. I was direct and unmerciful. Sounds okay, but this doesn't paint the clearest picture. I did all these things as a first step…without trying to understand her. For instance, if she made a mistake (or what I thought was a mistake) with her words, I would immediately pounce. I was under the impression that I was helping her be more careful and caring with her words; yet, I wasn't being careful and caring towards her. I was using my words to prove her wrong, not to build her up.

We ought to be careful and caring with our words at every level of intimacy, especially the levels that require being loving towards one another.

In Matthew 12:36, Jesus said we will be judged for every idle word. What excuse will I have if I am not careful and caring with my words? What excuse would I have about my interactions with Morgan, a woman I was meant to be fellowshipping with? Doesn't not being careful and not being caring with my words directly result in my words being "idle"?

What does it mean to be careful and caring with my words?

Careful means I know the definitions of the words I use and how to apply those words.

- If I don't define the words I use, then others define them for me.
- I cannot justly get upset if they misinterpret my words if I don't define them.
- Words are important! They are the building blocks of our beliefs and our communication!

Caring applies to the intent of the words.

- Are my words for your benefit?
- Am I nourishing with my words? (presenting *truth*?)
- Am I cherishing with my words? (an environment of *love*?)
- Or do I tear people down in order to feel right?

Notice, fellowship allows me to initiate sharing on Morgan. In order to help Morgan feel safe, she was always able to ask me how her answering my question was a benefit to her. If I couldn't answer that question, then I wasn't really careful and caring with the intent of my words.

Despite all of this, the fellowship between Morgan and I drifted towards toxicity. I praise God that I started feeling unsettled and guilty. So I shared with her that we ought to move a step back into the relationship level of intimacy where neither of us was expected to speak into each others' lives nor share everything we HAVE, DO, and ARE with each other.

The levels of intimacy are measured by profitability. My behavior towards Morgan destroyed the profitability of our intimacy. When we stepped back into relationship, things became profitable again, and this is what allowed us to continue to grow in our intimacy, without abuse.

If all levels of intimacy are measured by profitability, fellowship is driven by love. Looking back to what I wrote above, it's clear there came a point where I stopped loving Morgan. I was often not even giving a value, and I had expectations. Morgan and I have repaired. We both now look back on that tumultuous time as such a great learning experience. We both are so appreciative of the tools we've learned since then that have allowed us to communicate in a more careful and caring manner with each other and the other people in our lives.

CHAPTER 3

Communication

BOOK 2 BEGAN with us stating the party itself is the ultimate in intimacy and communication by exchanging with others through our AREs. We looked at intimacy, so now it's time to look at communication.

Communication is defined as the exchange of information. Since our measure for everything is profitability, we need to look at communication relative to profitability, which can also be distinguished as healthy and unhealthy communication.

There are three basic guidelines for healthy and profitable communication.

1. Make statements on yourself
2. Ask questions of others
3. Answer other people's questions

Making statements *on* yourself is beginning sentences with: I think, I feel, I believe, It seems to me, etc. These statements are intimate. This is different from making statements *about* yourself which is like bragging: "I won the tournament."

The opposite of these three guidelines are unhealthy and unprofitable communication.

1. Make statements on others
2. Don't ask questions of others
3. Don't answer other people's questions

My favorite way to illustrate this is with a story from God's word that almost every Christian believes they know so well, they are almost bored reading it.

> *And they heard the voice of the Lord God walking in the garden in the cool of the day: and Adam and his wife hid themselves from the presence of the Lord God amongst the trees of the garden. And the Lord God called unto Adam, and said unto him,...* (Genesis 3:8-9a)

What did God say to Adam? Take a moment to write or state out loud what you believe God said before continuing.

> *And the Lord God called unto Adam, and said unto him, Where art thou? And he said, I heard thy voice in the garden, and I was afraid, because I was naked; and I hid myself. And he said,...* (Genesis 3:9-11a)

God asked a question: Where are you? Adam answered the question, and then what did God say? Again, take a moment to write or state out loud what you believe He said before continuing.

While you are thinking about that, here's a question: Why did God ask Adam where he was? Did God not know where Adam was? While you think about that, let's see how God responded to Adam.

> *And he said, Who told thee that thou wast naked? Hast thou eaten of the tree, whereof I commanded thee that thou shouldest not eat? And the man said,...* (Genesis 3:11-12a)

God responded with two questions! How did the man respond? While you think about that, here's another question: Did God not know who told Adam he was naked and that Adam had eaten of the tree? While you think about that, let's see how Adam responded to God.

> *And the man said, The woman whom thou gavest to be with me, she gave me of the tree, and I did eat.* (Genesis 3:12)

Did Adam answer either of God's questions? He didn't answer the first question and while he did answer the second question, his answer seemed to imply it was God's fault (*The woman whom thou gavest to be with me*). How did God respond to Adam?

While you are thinking about that, clearly, God knew the answers to all three questions He asked Adam; otherwise, He isn't God. Clearly, God didn't ask Adam these questions because He didn't have the information or do you think God doesn't have all the information?

God wasn't trying to see where Adam was at physically. God was trying to see where Adam was at mentally, emotionally, and spiritually. God was giving Adam the opportunity to admit what he did was wrong, and Adam proved where he was at by not answering some questions and blaming God as his answer to other questions.

> *And the Lord God said unto the woman, What is this that thou hast done? And the woman said,...* (Genesis 3:13a)

God didn't respond to Adam after Adam blamed God. God put His focus on the woman and asked her a question. How did the woman respond? While you are thinking about that, notice that so far, God hasn't made one statement! How many of your answers for God in this passage were a statement?

> *And the woman said, The serpent beguiled me, and I did eat.* (Genesis 3:13b)

The woman did answer the question; however, she didn't blame God as Adam did, she blamed the serpent. How did God respond to this? While you are thinking about that, do you think God didn't know what the woman had done? As you already realize with the man, God was finding out where the woman was at mentally, emotionally, and spiritually.

> *And the Lord God said unto the serpent, Because thou hast done this, thou art cursed above all cattle, and above every beast of the field; upon thy belly shalt thou go, and dust shalt thou eat all the days of thy life: And I will put enmity between thee and the woman, and between thy seed and her seed; it shall bruise thy head, and thou shalt bruise his heel.* (Genesis 3:14-15)

God didn't respond to the woman. God immediately turned His focus onto the serpent and God didn't ask a question. God made a statement of judgment. In fact, God followed this up by cursing the woman and the man (Genesis 3:16-19). Then God said this:

> *And the Lord God said, Behold, the man is become as one of us, to know good and evil: and now, lest he put forth his hand, and take also of the tree of life, and eat, and live for ever: Therefore the Lord God sent him forth from the garden of Eden, to till the ground from whence he was taken.* (Genesis 3:22-23)

What does that sound like? Notice, we have seen this ended the first dispensation and we see two judgments. One resembled a salvation judgment in that the man and woman were driven from God's presence. The other judgment resembled a reward judgment in that their profitability was hindered through curses.

It sounds as if God had a council meeting: *Behold, the man is become as one of us...*

Communication Causes

The second way to measure the profitability of communication is to look at the causes of the communication. Another way of saying this is looking at how the communication began; that is, what was the cause of the communication? There are four causes.

Good: Open-ended question or statement of fact.
Examples: "How was your day?" and "My team lost."

Not Bad: Close-ended question or statement of your opinion. "Did you have a good day?" and "I think my team should have won."

Bad: Projection (telling someone what they think or feel) or judgment. "You didn't have a good day." and "You're stupid for not liking my team."

Worst: Negate another person.
"What is your favorite color?" (They answer "blue".)
You state: "No, it's not. It's pink."

Notice, good and not bad causes give up control by giving the other person freedom to answer. Bad and worst causes attempt to exert control. In fact, when a female is given a good or not bad cause, she tends to share. When a female is given a bad or worst cause, she tends to shut down. However, when a male is given a bad or worst cause he tends to ramp up. In fact, almost every male who ends up in jail is a male who responded to a bad or worst cause with a bad or worst cause.

It turns out, you can determine a person's thought process by their communication and we've seen how these thought processes correlate to the levels on the wall.

> **Evil/Destructive:** Bad or worst causes. It doesn't matter if they use the communication guidelines or not. For example: "Why are you an idiot?" is an example of unprofitable communication even though it is a question. If it's unprofitable, it's evil and destructive.
>
> **Animal/Flesh:** Good or not bad causes stated in opposition to the healthy communication guidelines. For example: "You have freckles." While this is a statement of fact (good cause) it is a statement on another person. What is the result? The person begins to wonder: Is it wrong to have freckles? Why did you say that to me when other people have freckles?
>
> **Human/Man:** Good or not bad causes stated according to the healthy communication guidelines. For example, the above statement could be changed to "I see you have freckles."
>
> **Mada:** Conjunctive because it is truth. For example, Jesus said, *He that is without sin among you, let him cast first a stone at her* (John 8:7b).

Bullies

A bully is a person who is seeking to harm, intimidate, or coerce someone perceived as vulnerable. The definition of intimidation is "to make timid or fearful; frighten; especially, to compel or deter by or as if by threats" (Merriam-Webster).

Basically, a bully is looking to take a person from a human thought process to an evil thought process. The way they do this is first to overwhelm the person, which moves them from human to animal. This can be done with a question that is meant to overwhelm the person. Now the person is one threat away from going to an evil thought process. This is usually done with a statement.

The easiest way to determine if someone is a bully is to see if they follow the unhealthy communication guidelines. All bullies make statements on others and don't answer questions. When they do ask questions, it won't be according to good or not bad causes. Their bad and worst cause questions are meant to trap you. In addition, bad and worst causes are the definition of abuse. Abuse is interacting in a manner that brings about a bad effect or purpose; to unprofitability.

What's the simplest way to deal with a bully? Ask questions!

> *But when the Pharisees had heard that he had put the Sadducees to silence, they were gathered together. Then one of them, which was a lawyer, asked him a question, tempting him, and saying, Master, which is the great commandment in the law? Jesus said unto him, Thou shalt love the Lord thy God with all thy heart, and with all thy soul, and with all thy mind. This is the first and great commandment. And the second is like unto it, Thou shalt love thy neighbour as thyself. On*

these two commandments hang all the law and the prophets. (Matthew 22:34-40)

The Pharisees tried to trap Jesus by asking what one commandment is the greatest, and Jesus answered with two commandments because one was the cause and the other was the effect. Once Jesus answered their question that was intended to be unprofitable, Jesus responded with a question.

> *While the Pharisees were gathered together, Jesus asked them, Saying, What think ye of Christ? whose son is he? They say unto him, The son of David. He saith unto them, How then doth David in spirit call him Lord, saying, The Lord said unto my Lord, Sit thou on my right hand, till I make thine enemies thy footstool? If David then call him Lord, how is he his son? And no man was able to answer him a word, neither durst any man from that day forth ask him any more questions.* (Matthew 22:41-46)

They didn't want to answer the question because they would be proven wrong. If they wanted to grow and learn, they would have taken a shot at answering the question and seeing how close to the bull's eye they would have gotten. Their avoiding answering proved what was in their heart. That's how a bully works. They get you to talk and then project contradictions onto your words, which is the reason they don't talk: they don't want you to do to them what they are doing to you. What do you think of this next example from Jesus?

> *Then Pilate entered into the judgment hall again, and called Jesus, and said unto him, Art thou the King of the Jews? Jesus answered him, Sayest thou this thing of thyself, or did others tell it thee of me? Pilate answered, Am I a Jew? Thine own nation and the chief priests have delivered thee unto me: what hast thou done?*

> *Jesus answered, My kingdom is not of this world: if my kingdom were of this world, then would my servants fight, that I should not be delivered to the Jews: but now is my kingdom not from hence. Pilate therefore said unto him, Art thou a king then? Jesus answered, Thou sayest that I am a king. To this end was I born, and for this cause came I into the world, that I should bear witness unto the truth. Every one that is of the truth heareth my voice. Pilate saith unto him, What is truth? And when he had said this, he went out again unto the Jews, and saith unto them, I find in him no fault at all. (John 18:33-38)*

Jesus couldn't say, "I am the Son of God" because even though it was a right *what*, the *why* and *how* would be wrong. He would be doing it for His own benefit. Remember, truth is a right *what* with a right *why/how*. Jesus said as much in John 5:31 with:

> *If I bear witness of myself, my witness is not true.*

Jesus answered every question of Pilate's except the last one. Why didn't Jesus answer the question?

In the same way Jesus couldn't facilitate His own purpose and progress by testifying of Himself, He didn't have to answer a question that would benefit Him. If Jesus had explained the definition of truth, Pilate may have realized Jesus' answers were as close as Jesus could get to stating He was the Son of God and King of the Jews, and then He wouldn't have crucified Him. There is one more example of Jesus not answering a question.

> *And it came to pass, that on one of those days, as he taught the people in the temple, and preached the gospel, the chief priests and the scribes came upon him with the elders, And spake unto him, saying, Tell us, by what authority doest thou these things?*

or who is he that gave thee this authority? And he answered and said unto them, I will also ask you one thing; and answer me: The baptism of John, was it from heaven, or of men? And they reasoned with themselves, saying, If we shall say, From heaven; he will say, Why then believed ye him not? But and if we say, Of men; all the people will stone us: for they be persuaded that John was a prophet. And they answered, that they could not tell whence it was. And Jesus said unto them, Neither tell I you by what authority I do these things. (Luke 20:1-8)

Again, the question from the religious leaders was a trap. Jesus' answer would have been that He got His authority from God, the same as John the Baptist. However, Jesus dissolved the issue by asking the Pharisees their answer as to where John got his authority. This question from Jesus was mada because it quickly exposed the religious leaders' thought processes and their contradiction. When they passed on answering Jesus, it was right and just of Jesus *not* to answer them. If they had answered the question, I believe Jesus would have answered their question because that is right and just.

Summary

We have seen that we can measure the profitability of a relationship by the level of intimacy and the communication. In fact, it looks like we can simplify this by saying the health of a relationship = Sharing - Abuse.

For example, if two people are in fellowship and they periodically make statements on each other, the relationship is going to last because fellowship is a high level of sharing, while a relatively small number of statements is a low level of abuse. On the other hand, two coworkers at connection are going to have an unprofitable effect on the company if one or both are judging and negating each other because this is a low level of sharing with a high level of abuse.

Are you helping or hindering God's will with respect to communication?

In order to complete our equation, we need to cover one more measure of profitability.

Dr. Joel Swokowski's Commentary

We refer to these communication guidelines as guidelines for a reason. They work in most contexts but are not foolproof. We add the four causes along the three guidelines to help even further, yet, they can still be misused. As much a benefit as these guidelines are, paired with good and not bad causes, the intent of the communicator plays a vital role: are they looking to love the person or benefit themselves?

Again, as much a benefit as the guidelines are, there's an equal benefit in having the measures for when a person is using destructive communication and bad and worse causes. Here are some contrasting examples from the scriptures that bring this to light.

After the Pharisees said that Jesus cast out demons by the power of Beelzebub, Jesus stated the following,

> O generation of vipers, how can ye, being evil, speak good things? for out of the abundance of the heart the mouth speaketh. (Matthew 12:34)

Although Jesus asked a question, it could be seen as a bad cause if you see "being evil" as a judgment. This could also be seen as a fact.

Remember, Jesus only did what the Father taught Him to do (mada), so I choose not to use Jesus as the excuse for me making judgments on others. Furthermore, this was in response to the Pharisees judging Him, which would be evil. Jesus didn't make any judgment without the Father, nor did Jesus initiate it.

The story of Jesus and Legion is recorded in Mark 5. It is the story of a man who "always, night and day, was in the mountains, and in the tombs, crying, and cutting himself with stones."

The story continues:

> *But when he saw Jesus afar off, he ran and worshipped him, And cried with a loud voice, and said, What have I to do with thee, Jesus, thou Son of the most high God? I adjure thee by God, that thou torment me not. For he said unto him, Come out of the man, thou unclean spirit.* (Mark 5:6-8)

Here, to an evil spirit, Jesus directly calls the spirit unclean. Jesus wasn't using the communication guidelines when speaking to the evil spirit. In fact, whenever Jesus spoke to the spiritual realm, He used the opposite of the communication guidelines, and it always worked.

Now, for a contrasting example, here is someone using the communication guidelines to the spiritual realm. Let's see if it works.

> *And God wrought special miracles by the hands of Paul: So that from his body were brought unto the sick handkerchiefs or aprons, and the diseases departed from them, and the evil spirits went out of them. Then certain of the vagabond Jews, exorcists, took upon them to call over them which had evil spirits the name of the Lord Jesus, saying, We adjure you by Jesus whom Paul*

> *preacheth. And there were seven sons of one Sceva, a Jew, and chief of the priests, which did so. And the evil spirit answered and said, Jesus I know, and Paul I know; but who are ye? And the man in whom the evil spirit was leaped on them, and overcame them, and prevailed against them, so that they fled out of that house naked and wounded.* (Acts 19:11-16)

Here we see the sons of Sceva, a priest for that matter, did use the communication guidelines, and the evil spirit made a mockery of them and beat them while using the communication guidelines for himself. I can tell you, Lenhart is going to do a deeper dive into this in the third and final edition. I wanted to take this moment to clear up any confusion you may have applying these guidelines to the Bible.

For now, we can say:

The guidelines give us measures for when a person is bringing a benefit to another person, and the guidelines give us measures for when a person is being a bully to another person.

These measures ought to bring a person back to determining whether they're communicating in love or not.

As for the spiritual realm, it looks as if using the opposite of the communication guidelines is the effective way to speak, which begs the question: Was Jesus talking to the spiritual influence behind the Pharisees when it looked as if He was judging them?

CHAPTER 4

Repair

The thief cometh not, but for to steal, and to kill, and to destroy: I am come that they might have life, and that they might have it more abundantly. (John 10:10)

JESUS BLATANTLY STATED the enemy is focused on unprofitability, while He came that we may have life and life abundantly. In *Modeling God*, we saw the definition of life is "the ability to repair."

Unfortunately, we are going to hurt and abuse each other. As we saw at the end of the previous chapter, if we keep the abuse low while increasing the sharing, we can have a profitable relationship. However, our goal is maximum profitability, which means we ought to be striving for fellowship and eliminating abuse, but we will never eliminate abuse completely. We saw in *Modeling God* that we will never eliminate sin, but that is okay because God doesn't judge our sin, He judges our response to sin. Likewise, repair is our response to abuse.

Repair is an exchange (right *what*) that takes an unprofitable situation and makes it profitable through sharing (right *how/why*). Repair is ultra-sharing.

For example, if my sharing is at a +40 and my abuse is at a -30, then the profitability of my exchange with this person is +10. However, if I can repair the -30 to +30, my profitability goes from +10 to +70!

Notice, if people aren't profitable at sharing, it is going to be much harder for them to be profitable at repairing because it is ultra-sharing. What is this process?

Repair Process

Joel covered the repair process in his *Modeling God* commentary.

When we're dealing with repairing the issues we have with people, even the spiritual, emotional, and mental issues, we can take the "bad" things that happen and repair them. Repair with people comes through confession and repentance. I see repentance as repair that is done as an effect of having done something wrong. This helps us understand the measure for repair: when the people involved are happy that the "bad" thing happened because now that they have repaired, their relationship is better than it was before the "bad" thing. This means that when we handle the issues we have with each other according to God's instruction (confess & repent), the situation can be made good.

What if the person who did the abuse doesn't think they abused someone? How do we repair that situation? There is a three-step process to initiate repair.

1. **Step IN — Both parties recognize the issue.**
   This is similar to interaction in the levels of intimacy. We are only looking for both parties to be aware there is an issue. This

is tense and confrontational. The opposite of stepping IN is denial or avoidance.

2. Step UP — Both parties understand the other party.

This is similar to connection in the levels of intimacy. The focus is on understanding, not agreement. Understanding is an intentional cause that is in the control of both individuals. Agreement is out of the control of the individuals; you can't intentionally agree if you don't actually agree. This step is completed when both parties can state the position of the other party to the other party's satisfaction. Notice, Matthew 18 dealt with understanding, not agreement. Agreement is an effect of understanding. Focusing on agreement is the same as treating an effect as a cause.

3. Step THROUGH — Both parties reach a conclusion.

This is similar to relationship in the levels of intimacy because we are going to reach a conclusion on how both parties are going to relate to each other from this point forward. This is completed when both parties agree on their response to the issue. The opposite of this is to leave the issue open to be addressed later. Agreeing to disagree is stepping THROUGH as long as both parties have reached and individually stated their definite conclusions.

What about forgiveness? How does repair relate to forgiveness?

Forgiveness

Here's a tough question: Are we supposed to forgive everyone instantly? People are confused when it comes to forgiveness because they don't know how to deal with three seemingly contradictory passages.

> *Ye have heard that it hath been said, An eye for an eye, and a tooth for a tooth: But I say unto you, That ye resist not evil: but whosoever shall smite thee on thy right cheek, turn to him the other also. And if any man will sue thee at the law, and take away thy coat, let him have thy cloak also. And whosoever shall compel thee to go a mile, go with him twain. Give to him that asketh thee, and from him that would borrow of thee turn not thou away. Ye have heard that it hath been said, Thou shalt love thy neighbour, and hate thine enemy. But I say unto you, Love your enemies, bless them that curse you, do good to them that hate you, and pray for them which despitefully use you, and persecute you; That ye may be the children of your Father which is in heaven: for he maketh his sun to rise on the evil and on the good, and sendeth rain on the just and on the unjust. For if ye love them which love you, what reward have ye? do not even the publicans the same? And if ye salute your brethren only, what do ye more than others? do not even the publicans so? Be ye therefore perfect, even as your Father which is in heaven is perfect.* (Matthew 5:38-48)

Jesus stated not only should we forgive, we should be willing to help the person continue to be unjust to us!

> *Moreover if thy brother shall trespass against thee, go and tell him his fault between thee and him alone: if he shall hear thee, thou hast gained thy brother. But if he will not hear thee, then take with thee one or two more, that in the mouth of two or three witnesses every word may be established. And if he shall neglect to hear them, tell it unto the church: but if he neglect to hear the church, let him be unto thee as an heathen man and a publican.* (Matthew 18:15-17)

Jesus stated we ought to confront people who are unjust to us and if we are not heard, continue to get witnesses until we consider the person not a believer.

> *Take heed to yourselves: If thy brother trespass against thee, rebuke him; and if he repent, forgive him.* (Luke 17:3)

Jesus said we should confront people who sin and then forgive them when they repent!

Are we supposed to forgive people who abuse us, confront them so they listen, or only forgive when they repent?

The key to understanding each passage is to ask this question first: Is the sinner a believer or not?

Passage #1: If the sinner is not a believer, forgive him immediately in order to get more reward at his expense. Don't Step IN.

Passage #2: If the sinner is a believer, confront him (Step IN) so he will listen. This is a benefit to the sinner's conscience. If he doesn't listen (Step UP), then consider him to be an unbeliever.

Passage #3: If the sinner is a believer and he listens (Step UP), but he doesn't repent (Step THROUGH), extend mercy to him, which is also love. Don't forgive him because you are in community with this believer and any judgment he gets now may affect you. If the person never repents, then it will be solved on Judgment Day. Realize the person may not have sinned. We could be wrong and we will find that out on Judgment Day.

Look again more closely at the Luke 17:3 passage. Notice, this passage did not say it was a sin to forgive the unrepentant believing sinner. It stated if the person repents (repairs), that is, takes an action to make up for the injustice, then you can safely forgive them because there isn't any action God would have to take to equal out justice. According to Jesus, extending mercy until repentance occurs is how we ought to deal with believers who we believe have sinned against us.

However, if you are confused by the situation and want to play it safe you could forgive believers without their repentance. Realize you are actually not being merciful. You would be treating believers as unbelievers and encouraging God to take action against people, possibly without them understanding why they got a sudden punishment from God. In fact, if the reason you forgave was so you didn't have to confront the believer (Step IN), you would be in the wrong because you would be ignoring Jesus' command to confront the sinning believer.

While all three of these passages were ultimately guided by love, notice Jesus stated we have to confront believers and have to immediately forgive unbelievers. However, there are many churches reversing this. They encourage you to confront unbelievers in their sin and immediately forgive believers (without confronting them) in their sin. Not only is this damaging the person's thought process, it is hindering our ability to get reward. Worse, the person using this approach is hindering God's will.

God's Ability to Move

Jesus said the following to the church Laodicea in the Book of Revelation:

I know thy works, that thou art neither cold nor hot: I would thou wert cold or hot. So because thou art lukewarm, and neither hot nor cold, I will spue thee out of my mouth. (Revelation 3:15-16)

We have seen that people who are in God's will are gaining spiritual value (reward). These people were referred to as *hot*. We have seen that people who are hindering God's will are losing spiritual value. These people, who thought they were going *hot*, were referred to as *cold*. However, God's ability to accumulate value through justice at the *cold* person's expense allows God to respond to the person in order to get them back into God's will. We saw the way God does this is to increase the tension and we called it God's good will. God wants everyone to be in His perfect, pleasing, or good will so His eternal plan comes about quicker, and less people end up in the lake of fire.

Another way to look at this is the people who are *hot* or *cold* are stating their will through their works; they are definitely in or out of God's will and it is apparent to everyone. The only way to be anything other than *hot* or *cold* is to be in the process of deciding whether a person wants to be *hot* or *cold*. For example, we saw in Luke 17:3 that Jesus commanded believers to confront other believers that had abused them.

Imagine a believer complained that another believer had abused them and you asked them when they were going to Step IN to the confrontation. If the believer said they were still trying to decide if they actually got abused and continued to avoid making a decision, that person would be *lukewarm*. Notice, they would be better off stating they weren't abused and moving on with their life than remaining undecided because God can't help them. We saw in *Modeling God*, God would extend mercy in order to give them time to make up their mind. However, this is the ultimate way to hinder God's will.

Jesus stated that being *lukewarm* (not stating your will) causes Him to spit/vomit you out of His mouth. Our man-made tradition believes being *cold* and wrong is evil. We've seen that God knows how to repair abuse. Jesus stated in God's word that *lukewarm* is evil because it opposes God's will. Do you see *lukewarm* as opposing God's will?

We saw how God is accomplishing His will through everyone. I said it this way: God is accumulating $100 bills through those who are allowing God to do His will through them (*hot*), while God is accumulating pennies through those who are resisting God's will. Now we see that lukewarm people are the ones responsible for God only being able to accumulate pennies because they are delaying God's ability to gain any spiritual value. This means when we let people remain *lukewarm*, we have a share in delaying God's will. Notice, God will eventually accumulate spiritual value from this person, but it may only happen once they reject the good will of God.

We ought to help the individual go towards what they believe is right, believing it is *hot*, and then they ought to be willing to find out if it was *cold* or *hot* based on God's response. Let me be clear, I'm not saying this is an excuse to willfully do something wrong. I ought to strive to do the right thing in all situations. Yet I shouldn't worry so much about being right that I never make a decision or take action. Remember, even if my decision is wrong, God is just to show me and help me grow! Do you think it's worse to be wrong or *lukewarm*?

When it comes to repair, an apology is a man-made way of attempting to repair. Apologies are *lukewarm*. There are no apologies in the Bible. It was always confession and repentance.

When it comes to sin, refusing to state whether you believe something that you are doing is a sin or not is *lukewarm*. Saying you don't want to turn away from the sin is *cold*. Saying you want to turn away from the sin and actually turning away and repairing is true repentance; it is *hot*. Remember, God would rather we were *hot* or *cold*.

Notice, Jesus' desire is that we state our will one way or the other. He is focused on the *how/why*. His immediate goal is not that we don't make a mistake. That would be a focus on the *what*.

Our attempt to be perfect in our own strength often leads to us not doing anything because we are more focused on avoiding being wrong than doing what God says (being right), and that leads to us being *lukewarm*. Actually, He would rather that we did something we thought was right, even if it was wrong, because His focus is life. God would rather us go *hot* or *cold*; be experts at the ability to repair.

Today, the goal for "Christians" appears to be don't ever be wrong; be perfect (without flaw), just like the Pharisees. Jesus didn't come so that we would be without flaw; He knows we aren't without flaw and our striving to never make a mistake is going to wear us out. It will also ultimately prevent us from going *hot*, from attempting to reflect His influence upon our heart (grace). Jesus is not focused on the *what*.

Jesus realized we are going to make mistakes. Jesus said we are going to experience damage. Jesus came that we would be able to repair and repair more abundantly. This requires repenting, creation, and profitability, which is focused on the *how/why*.

We've seen that people change for two reasons: achieve gain or fear of loss. Very few people change in order to achieve gain. These people are excellent. These people are growth focused because growth is the achieve gain principle. Growth is uncomfortable, so our flesh wants to be content with what we currently have. *Lukewarm* believes we are good enough.

However, everyone changes with fear of loss. The Bible is proof of this. God always approached people first with an achieve gain mentality, with a principle, with a benefit, with a cause. When people chose not to respond, God went to fear of loss with a law, a penalty, or a negative effect. Either way, you will be uncomfortable, but you choose whether it is to achieve gain or prevent loss.

Lukewarm doesn't result in growth. Going *hot* or *cold* results in growth. Do we currently resemble the church of Laodicea?

It looks as if the current church's beliefs are not based on the word of God. They are based on contradictory, man-made traditions because these "Christians" are trying to remain comfortable.

Are you getting people to state their will in order to facilitate God's will, or are you enabling people to stay *lukewarm*, hindering God's will?

Dr. Joel Swokowski's Commentary

In the journey of learning, living, and teaching God's truth from within *Modeling God*, I've had quite the experience with many pastors and religious authorities in this area. Part of that journey was me being invited to attend a meeting from the very organization, the Fox Cities Evangelical Ministers Fellowship (FCEMF), that gathered together to slander, bully, and mob John Lenhart (author of *Modeling God*).

I had no idea what was happening at the time, but I did believe it was God who set this situation up. The last thing I wanted to do was attend a meeting with the organization that not only attempted to destroy Lenhart, but my entire community. I can only guess that thirteen years after the abuse, the people in charge of the FCEMF forgot about me.

I put my all into the meeting and continued attending every monthly meeting. After a year and a half, I was invited to be an officer on their executive committee. Again, another invitation I accepted. Again, nothing I initiated.

After that first year as an officer, I served as the president. Seriously, I'm still dumbfounded by how this all came about, and I am edified every time I think about it. God's hand was so clearly involved in this situation I struggle to attribute it to anything else. During my year serving as the president, I put my focus on setting up the FCEMF for the next year, including preparing the president who would take over when my year was over.

By this time, it had been a few years of serving at these meetings and discussing the issues that were hindering the flow of the Holy Spirit in the region the FCEMF serves. This resulted in everyone involved knowing my connection to Lenhart. It also meant I had been discussing how the defamation issue from 2009 could be repaired from the inside of the FCEMF. What I found was the same behavior from these religious people as there was thirteen years before. One of the other officers had successfully pitted me against the rest of the executive committee.

This all happened while still trying to serve as president and replicate myself into the next president. I had a meeting with the woman who would take over, and I expressed my concern to her, saying, "As the president, I don't think you can be Switzerland anymore."

Up to this point, she had walked the line of never making a decision or plain statement over what she believed was right: the FCEMF or John Lenhart. She was *lukewarm*. She would refuse to state her belief and refused to read *Modeling God*. This was all fine and well when she served as a member in the FCEMF, but as she was about to step into the main leadership role as president, dealing with this still ever-present conflict would be her responsibility to navigate.

One thing she did that *was* right was to email me later, asking me what I meant by my Switzerland statement. Here's my response:

> *In my opinion, I've seen you operate with a "peace at all costs" mentality. Trying to get along with everyone. Trying to be on all sides of an argument.*
>
> *I believe this mentality is what led you to avoid getting the information available to you about what has happened in and through the FCEMF since 2009. I do believe you are trying to be objective but the opposite is happening: not having the information available causes people to be easily deceived.*
>
> *I used "Switzerland" to represent this "neutral" mentality. Unfortunately, this mentality has resulted in abuse being enabled.*
>
> *I don't believe God is a God of "neutrality"…God is a God of Truth. He stands for what is Right and Just. God has ALL the information which allows Him to be objective.*
>
> *The FCEMF voted you in to be the [President] for 2023. A leader is meant to facilitate the purpose and progress of others. A leader is meant to put the best interest of others ahead of their own. A leader needs as much information as possible about the context in which they are leading in order to lead effectively. A leader sides with TRUTH, regardless of the cost to them personally.*

Notice the communication used. It was healthy communication made up of good and not bad causes. All of it was in response to a question she asked.

I wanted what was best for her and the FCEMF after my time as president was over. I knew this wouldn't happen if the conflict was ignored. I knew this would only happen if the president was *hot* or *cold*, as Jesus instructed.

I found out the result of this exchange was her telling other people that I had given her an ultimatum; that she had to choose between the FCEMF or the church that I'm leading. Go ahead and reread what I wrote to her. Where did I make this ultimatum?

While it seems I could make the case that she was bearing false witness against me, at least I could be satisfied facilitating her towards *cold* with her decision to bear false witness against me.

CHAPTER 5

Transformation

And be not conformed to this world: but be ye transformed by the renewing of your mind, that ye may prove what is that good, and acceptable, and perfect, will of God. (Romans 12:2)

WE SAW THAT everyone is either progressing towards the evil thought process or the godly thought process. You can't stay in the human or animal thought process indefinitely. Being transformed is growing in mada, the godly thought process. Those who are transformed are able to prove if something is in the will of God or not.

It would seem that knowing if you are in God's will or against God's will would be the most important ability each person ought to crave. How many people do you know who are transformed? How do we detect if we are in God's will or against God's will?

The first answer is whether you are pursuing your own plan or God's plan for your life. What was your plan for your life? If you can't identify it, then you are probably still pursuing it. If you are able to state *your* plan for your life, and it settles you, then how and when did you officially throw off that plan and never pick it back up so that you could focus on God's plan for your life?

Second, what are you pursuing, and which thought process did it correspond with for the happiness question in chapter 15? Actually, there is a simpler way to determine if you are in mada or not. Did you notice what were the main differences from mada and the other three thought processes?

The other three thought processes are looking outside of the individual for happiness and correspond to the three non-godly thought processes. Looking for something destructive to happen to others is evil. Desiring a situation lacking tension is animal. Striving to accomplish a reward is human. All of these are looking externally, and therefore to tangible things to make them happy.

We saw God made our brains to build a tolerance to what we tangibly experience, which requires us to desire more in order to exceed the new expectation and attain only the same previous feeling of happiness, so another attribute of those three thought processes is the quantitative. The three thought processes that oppose God's will are characterized by the tangible and quantitative. Basically, we want stuff, and we want more of it.

> *Take ye therefore good heed unto yourselves; for ye saw no manner of similitude on the day that the Lord spake unto you in Horeb out of the midst of the fire: Lest ye corrupt yourselves, and make you a graven image, the similitude of any figure, the likeness of male or female,* (Deuteronomy 4:15-16)

On the other hand, mada is characterized by the intangible and qualitative. The spiritual realm is intangible and God specifically didn't take a form so we wouldn't be confused by looking to the tangible when we looked to Him to make us happy. God is also focused on the qualitative, the experience. After all, His plan is a party where we get to know each other better qualitatively, while allowing more of who God created us to be to come out qualitatively.

Notice, *Modeling God* began with another version of the happiness question. I had you imagine a party where you could do anything for as long as you wanted. What did you imagine doing? Do you think you would never get tired of doing that in eternity?

Do you now realize the only thing you won't get tired of is fellowship and allowing more of your uniqueness to come out in response to every situation and activity? Remember, fellowship is growing in being deeply known and deeply knowing others.

As we saw, progress in mada would be measured in profitability as an effect of intimacy and communication. Any conception of eternity that looks to golf, video games, or any tangible activity for happiness, completely misunderstands God and is an example of a person who has not been transformed. Ideally, these activities ought to facilitate you towards growing in intimacy with others through healthy communication.

There is nothing wrong, in and of itself, with having a mansion (whether on earth or in the new Jerusalem). However, looking to a mansion as your source of happiness is what will prevent your ability to experience happiness in the long term.

What do you value now? Is it tangible and quantitative? Is it intangible and qualitative? Do you see the cause of an issue as tangible (person), or do you see the issue the way Jesus did, as coming from an intangible cause?

Ultimately, the best way to know if you are in God's will is to ask yourself: Are you gaining spiritual value for God's will?

How do we gain spiritual value for God's will? The rest of this book will focus on:

1. Showing you how you gain spiritual value for God's will.
2. Showing you how to determine if the group you are in is gaining spiritual value for God's will.
3. Showing you how to determine if others are gaining spiritual value for God's will, to their benefit or to their detriment.

The next seven chapters are going to give a detailed look at the two institutions that are not only God's eternal plan, but they were also created to bring about His universal will. These are the two largest parts of God's will! The rest of this chapter will serve as an introduction and overview to these two largest parts.

Church And Marriage

God is right and just in *what* He wants, *why* He wants it, and *how* He goes about bringing it to pass. We have seen the meaning of life is church and marriage, and God gave us church and marriage to bring about His universal will.

The Bible concluded with a happily ever after ending. What does that look like? Notice, we never see a movie or story that explains what "happily ever after" consists of, yet we know it will occur on the new earth for eternity. We don't see it as possible in movies because we currently believe that lack of fallenness is the cause of paradise.

Happily ever after means our happiness and joy ought to continually grow; otherwise, our brains will form a tolerance to the highest level of happiness we have experienced. In *Modeling God*, we saw how our happiness and joy could continually grow as an effect of the ARE of the individual. What does happily ever after look like between Jesus and the Bride?

We have seen that we are mental, emotional, spiritual, and physical beings. We are made to be integrated beings, where all these areas work together! It is not right for us to expect one of these areas to sustain us over the long term. You cannot drive your happiness by focusing solely on:

- your job (mental example)
- your hobbies (emotional example)
- prayer (spiritual example)
- having sex, doing drugs, exercising, or drinking (physical examples)

Whatever rush of energy you get from this activity is not sustainable because it doesn't involve the entire being. Eventually, that area of your life will return to the original level. Then, because this is not right, you will experience guilt. The guilt will take you to a lower level. Now, the focus becomes how to get rid of the guilt and attain the original level.

It is good to know how to repair and be able to attain the original level. However, if you were able to drive all areas to a higher level, this continuous growth would be sustainable.

A fear of loss approach would try to maintain this higher level until you did something unbalanced (unsustainable) and then returned back to the original level because of guilt, which would motivate you to grow to the higher level. An achieve gain approach would look to reach an even higher level without having to experience guilt. Which of these sound like the way believers will grow in eternity?

Jesus is married to the Bride, which is made up of all of the believers. Jesus is going to push His Bride in every area to another level. The Bride is going to experience more joy, more life, more energy, and more happiness. Then, Jesus is going to push His Bride in every area again, and the effect is going to be even more life, energy, happiness, and joy. Here is the meaning of life verse:

> *And he gave some, apostles; and some, prophets; and some, evangelists; and some, pastors and teachers; For the perfecting of the saints, for the work of the ministry, for the edifying of the body of Christ: Till we all come in the unity of the faith, and of the knowledge of the Son of God, unto a perfect man, unto the measure of the stature of the fulness of Christ: That we henceforth be no more children, tossed to and fro, and carried about with every wind of doctrine, by the sleight of men, and cunning craftiness, whereby they lie in wait to deceive; But speaking the truth in love, may grow up into him in all things, which is the head, even Christ: From whom the whole body fitly joined together and compacted by that which every joint supplieth, according to the effectual working in the measure of every part, maketh increase of the body unto the edifying of itself in love.*
> (Ephesians 4:11-16)

Our goal is for all of us to grow unto a perfect man, unto the measure of the stature of the fulness of Christ. This is achieve gain! Jesus is going to grow the Bride to Her maximum, which is an equal to Jesus. We saw this is called husbanding and Jesus is the Bride's Husband. This growth will be measured according to what we have previously covered: sharing, abuse, and repair.

Is the Bride going to be able to propel Jesus in the same way? If not, then why did God need to provide a Bride for His Son? Why did God go through all this trouble? What does eternity look like when the Bride attains the stature of Christ? Isn't it impossible for the Bride to attain the stature of Christ? Why would we even think that all of us working together could do something that seems impossible?

Remember Genesis 11:6 covering the tower of Babel? Nothing will be restrained from them which they imagine to do. Besides, marriage means we become one flesh.

> *For the husband is the head of the wife, even as Christ is the head of the church: and he is the saviour of the body.* (Ephesians 5:23)

> *And he is the head of the body, the church: who is the beginning, the firstborn from the dead; that in all things he might have the preeminence.* (Colossians 1:18)

Doesn't that make the Bride God? No! If anything, it allows the Bride the ability to have complete fellowship with God. Even if the Bride was able to become "God" in eternity, this would not make any individual cell of the Bride God. After all, there are trillions of cells in the human body. It is not hard to imagine trillions of people making up the Bride.

Also, if we remember systems thinking, then we know the profitability isn't in the individual person or cell, it is in the exchange between individuals and cells. That is why sharing and repair from abuse are the foundational measures that drive the profitability of church and marriage.

So, it looks like the church is a body of believers of one mind who are acting in their perfection, their maximum profitability. There is nothing that the Bride will be unable to do, even being able to propel Jesus to have even more life, energy, happiness, etc. Notice, this makes our fellowship with God the Father similar to a woman to her father-in-law. We will be God's Daughter-in-law.

Church is the ultimate community solely made up of people (even though God can be a part of it). The profitability is in the exchange, not the individual. Also, everyone ought to benefit, so this is currently a great way to generate spiritual value for God's will.

Generativity is an interaction where everyone gains so that there is more than what was originally present. This is why we are able to generate tremendous spiritual value (e.g., $100 bills) when we are allowing God to work through us. God isn't interested in sustainability; God is interested in generativity.

In *Modeling God*, we saw there are two things you can do with your seed: eat it slowly (sustainability) or eat the minimum so you can plant the rest and end up with more (generativity). This is another way you can tell if you are in God's will or not: Are your actions sustainable or generative?

The ultimate community is marriage because God is involved. Remember, when there are three members, there are seven interactions; six ways to be profitable involving God. The meaning of life is the church (one type of ultimate community) in the ultimate community (marriage covenant with Jesus and God the Father).

Everything about the meaning of life is focused on creation through community, which is an effect of people loving according to their uniqueness. The enemy's strategy is unprofitability through denying the uniqueness of the individual and destruction through isolation.

The enemy isolates us from God by denying that grace is the divine influence upon the heart, and its reflection in the life. Now we see the danger in believing and teaching the man-made traditional definition of grace as "unmerited favor." The enemy also isolates us from each other by focusing us on appearance. Now we see the danger in believing and teaching the man-made traditional metric of lack of sin. Neither of these doctrines create because they deny uniqueness and focus on more stuff, which includes appearance.

Notice, this explanation for the meaning of life says that God needed us in order to create a Bride for His Son. However, this doesn't mean that God was lacking without us. He could exist forever without it. However, God had a desire for something more and Jesus was willing to give His life to accomplish it.

Basically, the Bible is saying that eternity is church and marriage. Everyone is connected through fellowship. There are no possessions; no things. There are no solo activities. There are no hobbies. The Bible is saying focusing on those activities are something less than interacting in your ARE with others. Do you believe that?

Church and marriage are something that is generative and able to be perfect. This can go on forever for people who are able to feel good about themselves because they feel good about others. This is fellowship!

However, notice something: our tradition says that the key to happiness is HAVING (quantitative) and DOING (tangible). People state they would be happy if they HAD certain things (money, big house, nice car, etc.) and/or able to DO certain things (travel, be in charge, play golf all day, etc.). When society views HAVING and DOING as the meaning of life, then the real meaning of life becomes unattainable to the members who accept society's values.

Summary

The meaning of life according to the Bible is ARE, not HAVE, and not DO. The Bible believes the ARE is generative and able to be perfect because it is qualitative and intangible.

Man's perspective is: "the meaning of life is HAVE and DO, and the ARE is not sustainable or able to be perfect (without flaw)," which was Solomon's strategy.

How many times have you heard a pastor talk about looking forward to a DO or HAVE in heaven? "I can't wait to play golf in heaven!" "Wait until you see me in my Cadillac in heaven!" Are they helping or hindering the will of God?

The only parts of life that shouldn't be a waste of time are church and marriage.

Did that last statement bother you? Why?

The Bible said that the only things that give us pleasure are interacting with others in our ARE: church and marriage. The Bible said all stuff is unprofitable; it will burn.

However, man looks at the aspects of this life that are lacking (HAVE and DO) and tries to make them profitable. Man tries to make possessions and activities a source of joy. When man applies that perspective to church and marriage, they both become unprofitable.

Those who covet society's values will believe that church and marriage are unable to be paradise on earth. Said another way, those who believe church and marriage cannot be paradise on earth have put man-made tradition in place of the word of God.

Part II of this book will present the biggest ways people are unknowingly hindering God's will by embracing man-made tradition. This is the most uncomfortable part of any of the three editions. Please take your time.

These areas can be seen as roadblocks on the route God has planned for you to reach the party. Let's begin with marriage!

Dr. Joel Swokowski's Commentary

Church and marriage are the meaning of life. We'll continue to learn more about what this means in future chapters. Until then, allow me to pose these same words in a different manner. My friend Jonathan was the first to ask these questions to me and it just hit different:

> Are church and marriage the *meaning* of your life?
> Or is church and marriage *part* of your life?
>
> Is God the *meaning* of your life?
> Or is God *part* of your life?

Sustainable vs. Generative

People want things to last; yet, we seem not to have figured out how to make them last. The phrase, "Remember the good ole days?" exhibits this perfectly. It shows we want the good times to last, but we haven't figured out how to do so.

Sustainable means able to be maintained at a certain rate or level; able to be upheld or defended. This sounds well and good, until you realize that there are no "zero" events. A thing is either growing or dying. This means that even the things that are sustainable are only so for the short term.

The reality is that even things that are "not bad" and "not unprofitable" can be seen as resulting in something dying because it didn't intentionally help it grow. The death isn't coming directly from the "not bad" or "not unprofitable" stimuli; it's coming from the fact that all physical processes are running down, even microscopically. For example, the older I get, the more muscle loss I experience. I noticed that once I reached my forties, if I did not continue putting effort into growing my muscles, even if I was "not bad" or "not unprofitable" with my nutrition, my physical body was breaking down. Is the "not bad" nutrition the direct cause of my muscle loss? No. But my lack of growth in the muscular parts of my body does mean I am losing muscle, or essentially experiencing "muscle death." If I'm not growing (generative) my muscles, I will not sustain my current muscular build.

Generative means relating to or capable of production or reproduction; to beget (give life). This is God's plan for us: life-giving and profitable.

Goliath's Isolation

Our enemy (the devil) wants to destroy. He does this primarily through isolating us from one another. The story of Goliath illustrates this perfectly.

> *Now the Philistines gathered together their armies to battle, and were gathered together at Shochoh, which belongeth to Judah, and pitched between Shochoh and Azekah, in Ephesdammim. And Saul and the men of Israel were gathered together, and*

> *pitched by the valley of Elah, and set the battle in array against the Philistines. And the Philistines stood on a mountain on the one side, and Israel stood on a mountain on the other side: and there was a valley between them. And there went out a champion out of the camp of the Philistines, named Goliath, of Gath, whose height was six cubits and a span.* (I Samuel 17:1-4)

Goliath from Gath was a champion of the Philistines, their most skilled and feared warrior. He was approximately ten feet tall!

> *And he had an helmet of brass upon his head, and he was armed with a coat of mail; and the weight of the coat was five thousand shekels of brass. And he had greaves of brass upon his legs, and a target of brass between his shoulders. And the staff of his spear was like a weaver's beam; and his spear's head weighed six hundred shekels of iron: and one bearing a shield went before him.* (I Samuel 17:5-7)

Goliath wore about 200 lbs. of brass armor. The head of Goliath's spear alone weighed approximately 25 lbs. For simple comparison, a rifle used in the military today weighs less than 8 lbs. In other words, Goliath is huge in physical stature…a GIANT!!

> *And he stood and cried unto the armies of Israel, and said unto them, Why are ye come out to set your battle in array? am not I a Philistine, and ye servants to Saul? choose you a man for you, and let him come down to me. If he be able to fight with me, and to kill me, then will we be your servants: but if I prevail against him, and kill him, then shall ye be our servants, and serve us. And the Philistine said, I defy the armies of Israel this day; give me a man, that we may fight together.* (I Samuel 17:8-10)

This was a negotiation: whoever wins a fight between just these two soldiers would win the entire war. Goliath is exhibiting a masterclass in bullying, using fear and intimidation to lower the esteem of his opponent. What Goliath is doing here is causing each of the Israelite soldiers to imagine fighting Goliath one on one. This is what divides them as an army and turns them into just a group of disconnected individuals.

> *When Saul and all Israel heard those words of the Philistine, they were dismayed, and greatly afraid.* (I Samuel 17:11)

In this verse, dismayed (H2865 hatat) meant to be shattered or broken. The Israelite army was disconnected. A group of people can still just be a group of individuals, isolated from one another. His negotiation caused fear in each of them and caused them to forget about the strength they had together, as an army.

Is the enemy effectively using this strategy on you?

PART TWO

Roadblocks to God's Will

INSIDE

Chapter 6: Marriage 301

Chapter 7: Marriage Covenant 315

Chapter 8: Divorce 330

Chapter 9: Putting Away 343

Chapter 10: Divorce vs. Putting Away 357

Chapter 11: Church 375

Chapter 12: Church Leadership 393

Chapter 13: Women 417

Chapter 14: Men 432

Chapter 15: Bringing About God's Will 451

The Next Book 478

CHAPTER 6

Marriage

THESE NEXT FIVE chapters cover a topic that requires a large number of Bible verses to fill it in as a subsystem. It can be very easy to feel like you have found a contradiction in this explanation, but remember, the non-contradictory answer requires information from all five chapters. If you feel something is wrong, make a note and continue reading.

We saw God used marriage as an analogy for God's covenant with Israel and Judah. Let's look at what God's word says about marriage.

> *And Isaac brought her into his mother Sarah's tent, and took Rebekah, and she became his wife; and he loved her: and Isaac was comforted after his mother's death.* (Genesis 24:67)

> *If a man find a damsel that is a virgin, which is not betrothed, and lay hold on her, and lie with her, and they be found; Then the man that lay with her shall give unto the damsel's father fifty shekels of silver, and she shall be his wife; because he hath humbled her, he may not put her away all his days.* (Deuteronomy 22:28-29)

A virgin, not betrothed (engaged), who has intercourse with a man is married. The Bible said consensual sexual intercourse is marriage! Our tradition says marriage is a legal arrangement, a contract. Furthermore, our tradition says that marriage is what occurs at a wedding ceremony after the couple has been engaged (betrothed). What did the Bible say about weddings?

> *When thou art bidden of any man to a wedding, sit not down in the highest room; lest a more honourable man than thou be bidden of him;* (Luke 14:8)

> *And Jesus answered and spake unto them again by parables, and said, The kingdom of heaven is like unto a certain king, which made a marriage for his son, And sent forth his servants to call them that were bidden to the wedding: and they would not come. Again, he sent forth other servants, saying, Tell them which are bidden, Behold, I have prepared my dinner: my oxen and my fatlings are killed, and all things are ready: come unto the marriage.* (Matthew 22:1-4)

The Bible said the wedding is a feast! It occurs in a house with rooms. Oxen and fatlings were killed. Both the Old and New Testaments never mentioned a wedding ceremony, or a state recognized marriage. This is what we have today. Marriage and weddings were different in the Bible than today.

Biblical marriage and Biblical weddings were ordained by God. Where did our man-made concept of marriage and weddings come from?

Traditional Marriage

Clearly, our concept of marriage and wedding came after Biblical times. In fact, translating the Bible into English was forbidden as late as the 1300s because the state didn't want the common man marrying more than one wife. Marriage was able to occur without both the state and the church, and everyone knew it!

State recognized marriage didn't occur until the 1500s through The Marriage Ordinance of Geneva: "The dual requirements of state registration and church consecration to constitute marriage." John Calvin was "ruling" Geneva during this time with an approach that fit the fifth dispensation of the law instead of the sixth dispensation (e.g., adulterers were stoned), so the tradition of "Christian marriage" was the work of John Calvin and his Protestant colleagues!

As for the public ceremony, this was made law in England and Wales by Lord Hardwicke's Marriage Act 1753. This act required a formal ceremony of marriage with vows to curtail the practice of "fleet marriage." A fleet marriage was a clandestine marriage that got its name because they took place at Fleet Prison.

Under English law at that time, a marriage was recognized as valid if each spouse had simply expressed (to each other) an unconditional consent to their marriage (betrothed). No particular words were necessary, no clergyman or registrar needed to be present, no witnesses were required—just a vow. Notice that the Act of 1753 didn't require the church to be involved at all.

Most people married in church; however, since the church desired it, and the family and friends expected it. Vows could be exchanged by a boy as young as 14 years old and a girl as young as 12 years old. The church expected parental consent for those under the age of 21 years old.

Mini-summary of the origin of our traditional marriage and wedding:

John Calvin introduced the concept of government marriage to the Western World in the 1500s when he combined the previous (fifth) and current (sixth) dispensations.

The Marriage Act of 1753 introduced the concept of a public wedding ceremony to the Western World. These man-made concepts (traditions) were created relatively recently and more than 1500 years after the Bible!

Whose Will?

If a person says that God is against consensual sex before marriage, the only way this can be true is if the person is defining marriage as the current state-approved marriage that is based on man-made tradition. This person has put the man-made tradition above God's definition of marriage and wedding!

If a pastor or "Christian" author makes this claim, they are teaching tradition in place of God's word! They are representing the government, not God, in this important and very public doctrine. Teaching man-made doctrine in place of God's doctrine is what the Pharisees did.

Again, it isn't necessarily a sin if this person has been putting tradition in place of the word of God and didn't know it. However, like Jesus with the Pharisees, the issue of sin comes once the person is made aware they are putting tradition in place of the word of God (John 15:22).

What does your pastor say when you make them aware of this?

Are they going to actively work to justify themselves and their tradition in place of the word of God? Are they going to confess and repent? Are they going to admit they have been following a tradition and repair the damage they've done by resisting God's will?

Or are they going to be lukewarm and try to ignore this information and go along with the crowd? Every time the discussion of marriage and weddings occurs in their presence, they are stating their will to God whether they are helping or hindering His will.

Biblical Marriage

Today, we know the wedding as a ceremony where the bride and groom get married. What did the Bible say the wedding ought to be?

The wedding is a supper or a feast to which guests are invited. The guests are greeted by the bride and groom. During the feast, the bride and groom go into the bridal chamber (Hebrew: huppah) to have sexual intercourse. The best man stands outside the door and waits for the groom's voice to signify when the marriage is consummated. Once sexual intercourse is completed, the couple is declared married if there is proof the bride was a virgin. The proof is the bloodstained bedsheet. In the Old Testament, this cloth was presented to the bride's parents.

> *If any man take a wife, and go in unto her, and hate her, And give occasions of speech against her, and bring up an evil name upon her, and say, I took this woman, and when I came to her, I found her not a maid: Then shall the father of the damsel, and her mother, take and bring forth the tokens of the damsel's*

> *virginity unto the elders of the city in the gate: And the damsel's father shall say unto the elders, I gave my daughter unto this man to wife, and he hateth her; And, lo, he hath given occasions of speech against her, saying, I found not thy daughter a maid; and yet these are the tokens of my daughter's virginity. And they shall spread the cloth before the elders of the city.* (Deuteronomy 22:13-17)

This is God's view of marriage according to God's word. Some may say we have a different culture and God would change His view on this doctrine to justify us. Would God change other doctrine (e.g., sin) according to our culture, especially as it relates to one of the two institutions that are part of His eternal plan and used to bring about His universal will?

Some may wonder what difference it would make to God if we followed our culture instead of His word. What is the problem with not believing consensual sexual intercourse is marriage?

If someone has had intercourse and then got legally married to someone else, according to God's word, they are married to the first person and committing adultery with a person our tradition considers their wife!

Our culture (and religion?) would be encouraging an act that is adultery in God's eyes and then wondering why the marriage is failing. Everyone would be wondering why the marriage isn't profitable. It wouldn't be God's fault.

The couple has hindered the will and word of God through their tradition. They have made it of none effect (unprofitable). We will see later there are a lot of verses in the Bible telling us how to treat our spouses otherwise we won't be blessed. In this case, God would be responding depending on how the individuals He saw as married treated each other.

Some people consider this the way marriage and weddings were in ancient times only. However, our current traditional view of marriage and weddings is relatively new; even the New Testament doesn't really qualify as ancient times, or does it? Do these same people want to ignore the rest of the New Testament?

Even if we wanted to believe God no longer views marriage this way, why did God want marriage to be consensual sexual intercourse and a wedding to be a feast? What was God showing us? Let's go back to God's perspective.

God's Marriage Perspective

We saw that from the beginning of the Bible, God considered consensual sexual intercourse equal to marriage. This is not *lukewarm*. This leads to the most objective method for people to know for sure if they are married and not in adultery, because people in adultery would result in chaos.

Remember, the responsibility is on us, not God. If people aren't sure they are truly married it is going to lead to people being unprofitable. How can a man and a woman know for sure they are marrying someone who isn't already married? How can people make sure they aren't committing adultery?

God wanted them to focus on whether the person has had intercourse. The focus was on virginity. How do we prove virginity objectively?

If we need an objective measure, it has to be physical. One of the two people had to bear the proof of virginity. The proof had to occur the first time this sex had intercourse. The proof is the bloodstained sheet. Now take a good long look at the weddings described in the Bible. They are actually feasts for a bloodstained sheet!

The bloodstained sheet ought to be celebrated! It is proof that one of the members of the society is excellent. It is proof God's will and word were being followed and the culture would continue.

Why do we look at that bloodstained sheet as disgusting or repulsive or barbaric? Is this a focus on appearance? What are we valuing? Why do we look at the requirement of women being a virgin as demeaning or worse, unfair? Doesn't it require the woman to be more excellent than the man?

Is it because our tradition has blinded us to the truth God is trying to tell us? Do we no longer know the *why* behind God's word and will? Is it because we don't see women as more excellent? Is it because women today aren't more excellent? A recent study by the Guttmacher Institute found that 46% of American high school students have had sexual intercourse.

God's view of marriage and weddings led to profitability in society. The sheet proved the couple was married and not in adultery! The sheet objectively proved she wasn't married to someone else. Virginity definitely proves she isn't married to someone else.

When did this change? When did we give up on God's word and embrace a man-made tradition? When did we lose spiritual value and become unprofitable?

We saw that John Calvin (1509-1564) made marriage an institution of the state. At about the same time, Martin Luther (1483-1546) said not to punish husbands too fiercely for having affairs because it is a drive that is divinely ordained to men. Recall, when Martin Luther defined grace as "unmerited favor," it wrongly taught our salvation became something for which we are not responsible. In the same way and at the same time, Martin Luther stated man's response to his divinely ordained sexual drive is something he was not responsible for because that was consistent with his worldview.

Isn't this a clear example of someone turning a man-made perspective (men's God-given sexual drive) into a doctrine (not a sin)? Why didn't Martin Luther focus people on God's doctrine?

Meanwhile, Calvin took Martin Luther's worldview and added government control. His result was to put adulterers to death like in the previous dispensation of the law. God didn't put the law on us, a man did.

Why do people combine dispensations and bring the government into God's word?

Why did people embrace a contradictory interpretation of the Bible?

Clearly, we have let a traditional view of marriage replace God's word and will for this important and public doctrine, and we are reaping the current unprofitable results. Every time pastors, teachers, and "Christians" speak of marriage or weddings in this man-made traditional manner, they are publicly professing to be the same as the Pharisees.

Let's go deeper and understand God's doctrine for marriage and how it ought to be.

Marriage Doctrine

God meant for marriage to be community. There is a lot of discussion today about community. What is community?

Community means with unity. We define community as "three or more individuals accomplishing something different than they could individually." The *something* defines the type of community.

If the *something* is provide quick meals at a low price, then the community could be a fast-food restaurant. If the *something* is provide coffee, then the community could be a coffee shop.

The goal of the community is profitability (God's metric). Profitability is achieved by exchanges between unique individuals. The profitability of a community can be determined by summing all the exchanges that occur in the community, that is, the value created or consumed from every exchange.

So, the only way to grow a profitable community is to add people whose exchanges result in a net gain. This can be done intentionally if the new members are given the tools to intentionally do this.

If a community is just interested in increasing membership, they are likely to end up unprofitable if they don't give the new members tools for being profitable. This is also why if a community encourages exchanges that didn't previously occur (e.g., small groups in church), it is likely to suddenly make the entire community unprofitable if the members aren't equipped with the correct tools. Also, if the individual's uniqueness is denied through laws and tradition, it will take a tremendous amount of energy to sustain this community in the long term.

Usually, a group or organization asks people to sign up for duties or responsibilities. If the people haven't been taught who they ARE, the community objective, etc., then they may sign up for things based on time or money or guilt reasons. This will ultimately result in their being drained of energy and not being profitable to the rest of the group. They may even be limiting someone from occupying a role that would result not only in the profitability of an individual, but the group as well. The role of leadership is to help groups achieve the objective by putting people in the proper roles and giving them the tools to allow themselves and others to be profitable.

Marriage is a community: a husband, a wife, and God. This is the most efficient way to create value because God is involved in six of the seven exchanges.

Notice that church is a community that includes Jesus!

> *For where two or three are gathered together in my name, there am I in the midst of them.* (Matthew 18:20)

Church and marriage are God-given communities. The purpose of a community is to be profitable; to generate value. These are two methods God gave us to generate spiritual value in such a way that everyone gains spiritual value now, and it doesn't require us to handle something unjust in order to gain that value!

God created both marriage and church as a way for us to be profitable and generate the spiritual value needed to bring about God's will! Now we see the importance of following God's word: it is profitable and leads us to bring about God's will. Generativity!

How will you respond to people who are hindering God's will by defining marriage according to man's tradition?

Dr. Joel Swokowski's Commentary

Marriage vs Wedding

I've seen my friends Jonathan and Morgan go from just meeting, to dating, to engaged, to married. I was honored to be the person they asked to officiate their wedding. I remember the call when they gave me the news of their wedding date. After congratulating them, my first question was, "When are you cutting the covenant?"

It may seem crass and unromantic; yet, they didn't flinch. They had their answer immediately. Why? Because it was planned. Jonathan and Morgan (and our greater community) had understood the doctrine of marriage according to God for quite some time. They had a definitive difference in their understanding of a marriage versus a wedding. They knew their wedding was a celebration; a feast. They knew their marriage was caused by "cutting the covenant."

Jonathan and Morgan have spent their entire relationship, from dating to marriage, with the same goal: grow in deeply knowing each other. Many relationships begin this way but stop once they get married because getting married is the goal, which really means the wedding is the goal. This is a finish line mentality with the goal of a wedding. What happens after the wedding? There needs to be another goal set. If the goal from the start is to grow in deeply knowing each other, this can continue forever.

Jonathan and Morgan's wedding surely was a celebration. Those in attendance knew what was being celebrated because we had seen the example being set by this relationship before it became a marriage, and we knew what the future would hold as these two continued to grow. In fact, the day after their wedding, at church, I congratulated them in front of the congregation and endorsed them as a married couple that could be seen as an example and would be available to help other marriages.

How many marriages do you know that are an example for others to follow?

How many of those marriages were examples the moment they were married?

People think Jonathan and Morgan have been married much longer than they have. As of this writing, they've been married five years (together eight), and when people find out they've only been married five years, they're dumbfounded. How does this happen?

Jonathan and Morgan have been focused on the cause of getting to know each other deeply in every area longer than they've been married, and they still grow in that cause to the benefit of their greater community.

Laws of the Land?

One of the opposing arguments against the explanation that covenantal sexual intercourse is marriage that I've come across is: We're supposed to follow the laws of the land.

First, although this concept is in the scriptures (Romans 13:1-2), is it meant for us to follow the laws of the land at the expense of God's doctrine or His commands?

Generally speaking, laws tell you what you cannot do, not everything you ought to do.

It's not against the law of the land to embrace and believe God's definition for marriage. It may result in the government not recognizing the marriage, but that doesn't mean the husband and wife aren't married.

What's more important?

- God's definition for marriage (that doesn't break the laws of the land)?
- Man's definition for marriage (that can enable people to break God's laws)?

Samaritan Woman

One of the ways we know Jesus is the Son of God is that even 2000 years later, the things He said and did are still ahead of His time. The people Jesus interacted with and *how* He interacted with them is according to mada (godly thinking) and, therefore, still teaches us today. Another example is when Jesus interacted with a woman caught in adultery (John 8), which was mentioned earlier when learning about dissolve in Chapter 12. That story still teaches us today.

In the commentary to that same chapter, I dug deeper into the story of Jesus with the Samaritan woman (John 4). I mentioned that Jesus didn't immediately give that woman salvation, even after she asked, but instead told her to get her husband. She said the man she was with was not her husband. Now we see this chapter on marriage confirmed what we suspected: This was the second recorded incident of Jesus speaking with a woman in adultery.

It seems like Jesus showed mercy to everyone, even known adulterers. Does the church today show mercy to these people, shun them, or stone them as John Calvin enforced?

CHAPTER 7

Marriage Covenant

Here are the levels of intimacy before we covered covenants:

1. **Interaction** — superficial sharing in one area
2. **Connection** — deep sharing in one area
3. **Relationship** — deep sharing in at least one area and superficial sharing in other areas
4. **Fellowship** — deep sharing in all areas

The first verse in the New Testament that used the word fellowship was the Greek word metoche (G3352) which was defined "participation, i.e. intercourse: — fellowship." In fact, the following verse is used by some to apply to marriage even though it is covers all forms of fellowship:

> *Be ye not unequally yoked together with unbelievers: for what fellowship hath righteousness with unrighteousness? and what communion hath light with darkness?* (II Corinthians 6:14)

Now we can add the fifth level of intimacy in its final form.

5. **Marriage Covenant** — Fellowship plus sexual intercourse.

Fellowship is the key to God's plan for marriage. Consensual sexual intercourse covers all five steps of a covenant. Our tradition says marriage is a legal arrangement; a contract. Notice, another definition for contract is "to become small," while a covenant can grow indefinitely. This requires growing in sharing in all areas, even sexually.

Look again at how God cut the covenant with Abram and notice the sexual imagery:

> *And he said unto him, Take me an heifer of three years old, and a she goat of three years old, and a ram of three years old, and a turtledove, and a young pigeon. And he took unto him all these, and divided them in the midst, and laid each piece one against another: but the birds divided he not. And when the fowls came down upon the carcasses, Abram drove them away. And when the sun was going down, a deep sleep fell upon Abram; and, lo, an horror of great darkness fell upon him. And he said unto Abram, Know of a surety that thy seed shall be a stranger in a land that is not theirs, and shall serve them; and they shall afflict them four hundred years; And also that nation, whom they shall serve, will I judge: and afterward shall they come out with great substance. And thou shalt go to thy fathers in peace; thou shalt be buried in a good old age. But in the fourth generation they shall come hither again: for the iniquity of the Amorites is not yet full. And it came to pass, that, when the sun went down, and it was dark, behold a smoking furnace, and a burning lamp that passed between those pieces. In the same day the Lord made a covenant with Abram,* (Genesis 15:9-18a)

In the chapter on covenants, I stated:

This act was very intentional. Not only were all five parts of the covenant covered, but look at the imagery! Abram put bloody pieces of heifer, goat, and ram across from each other. Then there was a smoking furnace and a burning lamp that passed between those fur covered bloody pieces of flesh. (This aspect of a covenant was covered in the Strong's definition.) The covenant was consummated.

Strong's definition for covenant (H1285 berit): "from #1262 (in the sense of cutting [like #1254]); a compact (because made by passing between pieces of flesh): — confederacy, covenant, league."

Now we also see why God used a marriage analogy to illustrate His covenants with Israel and Judah. Let's look more closely at the marriage covenant.

Marriage

We have seen how a wedding celebrates the consummation of the covenant known as marriage; it is a feast for a bloodstained sheet.

According to the Bible, there are three ways a covenant can be created: sexual intercourse, vows, and/or exchanging tokens (jewelry, gifts, etc.). Likewise, notice that marriage isn't only sexual intercourse. Since marriage is a covenant, it can be created with vows or exchanges of tokens as well.

Today, people are realizing they are unable to fully enter into covenants because of soul ties (intense emotional and spiritual connection) with other people. Soul ties are formed the same way as covenants: intercourse, vows, or exchange of tokens. This is because soul ties are covenantal in God's eyes.

This means it is possible to be married before intercourse through vows or exchange of tokens. However, the marriage covenant isn't recognized as completed (consummated) until there is sharing, going through something, agreement, bloodshed, and death. In order to avoid confusion with our language, we are going to refer to vows or exchange of tokens as betrothed, with the completion (consummation) of the marriage covenant as being consensual sexual intercourse. Realize, the Bible sometimes refers to betrothed couples as married.

While it is possible to be betrothed before covenantal sexual intercourse, it is not possible to have covenantal sexual intercourse before marriage because when sexual intercourse follows the covenant model, it is marriage. However, not all sexual intercourse is marriage.

Effects Of Sexual Intercourse

When we use the covenant model, we see there are seven effects of sexual intercourse in the Bible. These effects are: marriage (monogamy), polygamy, concubinage, rape, incest, prostitution, and adultery. Every act of sexual intercourse mentioned in the Bible can be categorized using the covenant model. Let's look at each.

Marriage is covenantal sexual intercourse because all five aspects of a covenant are completely fulfilled. We saw this in a previous chapter with Isaac.

> *And Isaac brought her into his mother Sarah's tent, and took Rebekah, and she became his wife; and he loved her: and Isaac was comforted after his mother's death.* (Genesis 24:67)

Polygamy is covenantal sexual intercourse (marriage) because all five aspects are completely fulfilled. This means the first wife is not only aware of, but also in agreement with the other wife and willing to share her husband, like Jacob with Leah and Rachel. Another example is when Abram covenanted with Hagar. Sarai not only accepted it, but it was her idea!

> *Now Sarai Abram's wife bare him no children: and she had an handmaid, an Egyptian, whose name was Hagar. And Sarai said unto Abram, Behold now, the Lord hath restrained me from bearing: I pray thee, go in unto my maid; it may be that I may obtain children by her. And Abram hearkened to the voice of Sarai. And Sarai Abram's wife took Hagar her maid the Egyptian, after Abram had dwelt ten years in the land of Canaan, and gave her to her husband Abram to be his wife. (Genesis 16:1-3)*

Concubinage is not covenantal sexual intercourse because sharing is limited. The male shares only food and shelter with the woman. It appears the only men who had concubines in the Bible were men who had more than one wife. For example, did you know Abraham had concubines who had sons?

> *But unto the sons of the concubines, which Abraham had, Abraham gave gifts, and sent them away from Isaac his son, while he yet lived, eastward, unto the east country. (Genesis 25:6)*

We even have a polygamy/concubinage example: Solomon was warned not to take many wives because it would lead him astray.

> *And he had seven hundred wives, princesses, and three hundred concubines: and his wives turned away his heart. (I Kings 11:3)*

Rape is not covenantal sexual intercourse because agreement and sharing aren't fulfilled. For example, Shechem had intercourse with Dinah against her will.

> *And Dinah the daughter of Leah, which she bare unto Jacob, went out to see the daughters of the land. And when Shechem the son of Hamor the Hivite, prince of the country, saw her, he took her, and lay with her, and defiled her. And his soul clave unto Dinah the daughter of Jacob, and he loved the damsel, and spake kindly unto the damsel. And Shechem spake unto his father Hamor, saying, Get me this damsel to wife.* (Genesis 34:1-4)

Incest is not covenantal sexual intercourse when the law prevents agreement. This is why incest was able to be covenantal before the law for children of Adam and Eve, Noah's grandkids, Abram and Sarai, etc. Lot and his daughters weren't covenantal because Lot was completely unaware sexual intercourse occurred. He couldn't be in agreement. One could even argue that he was raped.

> *And Lot went up out of Zoar, and dwelt in the mountain, and his two daughters with him; for he feared to dwell in Zoar: and he dwelt in a cave, he and his two daughters. And the firstborn said unto the younger, Our father is old, and there is not a man in the earth to come in unto us after the manner of all the earth: Come, let us make our father drink wine, and we will lie with him, that we may preserve seed of our father. And they made their father drink wine that night: and the firstborn went in, and lay with her father; and he perceived not when she lay down, nor when she arose. And it came to pass on the morrow, that the firstborn said unto the younger, Behold, I lay yesternight with my father: let us make him drink wine this night also; and go thou in, and lie with him, that we may preserve seed of our father. And they made their father drink wine that night also: and the younger arose, and lay with him; and he perceived not when she lay down, nor when she arose. Thus were both the daughters of Lot with child by their father.* (Genesis 19:30-36)

An example of rape and incest was Amnon having intercourse with Tamar against her will.

> *And when she had brought them unto him to eat, he took hold of her, and said unto her, Come lie with me, my sister. And she answered him, Nay, my brother, do not force me; for no such thing ought to be done in Israel: do not thou this folly. And I, whither shall I cause my shame to go? and as for thee, thou shalt be as one of the fools in Israel. Now therefore, I pray thee, speak unto the king; for he will not withhold me from thee. Howbeit he would not hearken unto her voice: but, being stronger than she, forced her, and lay with her.* (II Samuel 13:11-14)

Prostitution is not covenantal sexual intercourse because sharing is limited; a price is set. For example, Judah and Tamar agreed on a price.

> *When Judah saw her, he thought her to be an harlot; because she had covered her face. And he turned unto her by the way, and said, Go to, I pray thee, let me come in unto thee; (for he knew not that she was his daughter in law.) And she said, What wilt thou give me, that thou mayest come in unto me? And he said, I will send thee a kid from the flock. And she said, Wilt thou give me a pledge, till thou send it? And he said, What pledge shall I give thee? And she said, Thy signet, and thy bracelets, and thy staff that is in thine hand. And he gave it her, and came in unto her, and she conceived by him.* (Genesis 38:15-18)

Adultery is not covenantal sexual intercourse because sharing is limited and there is a lack of agreement. The participants aren't sharing what they are doing with their current spouse(s) and they aren't free from other covenants to be able to get in agreement. For example, David's intercourse with Bathsheba was adultery because Bathsheba's covenant partner was unaware of it, and so were David's. In fact, David's desire to keep it a secret resulted in him having Bathsheba's husband killed.

> *And it came to pass in an eveningtide, that David arose from off his bed, and walked upon the roof of the king's house: and from the roof he saw a woman washing herself; and the woman was very beautiful to look upon. And David sent and enquired after the woman. And one said, Is not this Bathsheba, the daughter of Eliam, the wife of Uriah the Hittite? And David sent messengers, and took her; and she came in unto him, and he lay with her; for she was purified from her uncleanness: and she returned unto her house.* (II Samuel 11:2-4)

Fornication

In the Old Testament, the word fornication occurred four times, with fornications occurring once. There are two Hebrew words that cover these usages.

zanah (H2181) - "a primary root; to commit adultery; figuratively to commit idolatry."

taznuwth or taznuth (H8457) - "from #2181; harlotry; figuratively idolatry."

The definition of the word itself literally dealt with sex and figuratively with idolatry. Yet, all five usages of the word *by God* literally dealt with idolatry.

In the New Testament, fornicator was sometimes used literally to deal with sex, sometimes literally used to deal with idolatry, and sometimes it wasn't specified.

porneuo (G4203) - "to act the harlot; figuratively to practice idolatry"

porneia (G4202) - "from #4203; harlotry; figuratively idolatry"

In the New Testament idolator was defined as "image worshiper; to give one's self to an image." Idolatry means to worship an image. Another way of seeing this is to think of worship as giving one's self to. The root for fornicator is the same as for porn. Pornography is the worship of an image; whether it is ink on paper or colored lights on a screen, the thing the person gives themself to doesn't involve anything except an image.

Regardless of whether you believe fornication is first focused on worshiping other gods or sex, it looks like fornication covers both applications. Why? What is the link between idol worship and sex? Both are covenantal. It appears God sees worship and sex as being very similar.

Fornication is the forming of a covenant with an image. It is based on appearance. Remember, people are made in the image of God. Idols are images. Fornication is forming covenants with something we value more than God.

This results in two conclusions:

1. Every time appearance is made pre-eminent, it is fornication.
2. Fornication is the opposite of Biblical marriage, breaking the covenant.

Summary

Marriage is a covenant that can be created three ways: vows, exchange of tokens, and consensual sexual intercourse. The covenant can begin with vows or an exchange of tokens, so we are calling this being betrothed.

God sees covenantal sexual intercourse as marriage regardless of man-made tradition. If sex occurs between unmarried people and it is consensual, and limits on sharing aren't stated, then it is an intentional act that is a physical expression of covenant intentions.

Many marriages consist of the spouses going outside the marriage to get energy or spiritual value to prop up their marriage. God intended marriage to be a profitable community involving God so that it could generate spiritual value to bring about God's plan. This means covenants and sexual intercourse are intimate actions that ought to be progressed towards profitability.

Can we be profitable in a marriage covenant with just anyone?

Are there people we will be unprofitable in a marriage covenant with no matter what anyone does?

What if we find out we are in a covenant with someone who is in fornication, that is, in a covenant with an idol?

What should we do if the marriage covenant becomes unprofitable and hinders God's will?

Dr. Joel Swokowski's Commentary

A friend of mine named Alaina reached out to me as she was learning about the doctrine of marriage. Here's our conversation.

Alaina:

> Hi Pastor Joel! I was going through a restoration meeting last night and we discussed the story of Jacob and Leah in regards to marriage. I had a question… Did Jacob know it was Leah he was having sex with and not Rachel? If he thought it was Rachel, how was a covenant formed, as that to me does not sound consensual (covenantal) … more like hoodwinked! Just wondering, thanks in advance!

Joel:

> Hello!!…hoodwinked…I love it! And actually, this is a great question and the aforementioned hoodwinking actually helps prove this point!!

Marriage is consensual (covenantal) sexual intercourse. Conjunctive time:

— consensual/covenantal (limitation)
— sexual intercourse (freedom)

Both need to exist for it to be marriage. So, I can have sexual intercourse with *anyone*, but it's only marriage if it's consensual/covenantal.

Genesis 29 has the Jacob & Leah story:

> *And it came to pass in the evening, that he took Leah his daughter, and brought her to him; and he went in unto her. And Laban*

> *gave unto his daughter Leah Zilpah his maid for an handmaid. And it came to pass, that in the morning, behold, it was Leah: and he said to Laban, What is this thou hast done unto me? did not I serve with thee for Rachel? wherefore then hast thou beguiled me?* (Genesis 29:23-25)

Here was the hoodwinking: we see that Jacob and Leah had intercourse; however, was it covenantal? Seems like that's still up in the air! Jacob actually confronts Laban about the hoodwinking!

> *And Laban said, It must not be so done in our country, to give the younger before the firstborn. Fulfil her week, and we will give thee this also for the service which thou shalt serve with me yet seven other years. And Jacob did so, and fulfilled her week:* (Genesis 29:26-28a)

Laban explained his behavior and Jacob agreed to the conditions. Verse 28a proved that Jacob came into agreement with this making it both

— covenantal, and…
— sexual intercourse.

> *and he gave him Rachel his daughter to wife also. And Laban gave to Rachel his daughter Bilhah his handmaid to be her maid. And he went in also unto Rachel, and he loved also Rachel more than Leah, and served with him yet seven other years.* (Genesis 29:28b-30)

Finally, we can see that all parties fulfilled their ends of the agreement, and once again, we see an example of marriage being covenantal sexual intercourse:

> *And he gave him Rachel his daughter to wife also...And he went in also unto Rachel.*

The hoodwinking helped prove this definition! It became consensual/covenantal after the fact. Jacob had the right to either agree to the marriage or not, and he chose to agree.

Idolatry

Since the term idolatry has caused some confusion over its biblical usage and dictionary definition, allow me to remind you of the following as it relates to determining the definition of a word from *Modeling God* in the commentary of the 8. Grace chapter:

What's the process you use to determine a definition of a word? What Lenhart has done, and what every Biblical scholar ought to do, is to first go to the scriptures for a definition. This worked great with the definition of faith, as it was clearly stated in Hebrews 11:1. With grace, we see another step to take. Since this term isn't explicitly defined within the scriptures, we go next to a concordance, as with the definition for grace presented here. However, the ultimate measure for what the accurate definition is for any term is the principle of Non-Contradiction. This is true whether I find a definition in what the Bible says or I find a definition in dictionaries (which are made by men), or I hear a definition from another teacher or book. The definition must be Non-Contradictory for it to be at the "bottom rung" or "doctrine" level."

Lenhart's explanation above for idolatry is another example showing that it doesn't matter what a dictionary says if that definition contains contradictions. Lenhart used the dictionary from the original languages and showed how to understand this concept by being contrastive and removing contradictions. What's your measure for having a correct definition?

Profitable Marriages and the "No Touching" Strategy

One of the issues or concerns that people have is, "How can I know if I'm going to be profitable at the marriage covenant level with another person?" This is especially true for people who are dating and often brings an unsettled feeling to people who are already married.

The immediate answer is for these people to make sure they are in fellowship with each other before they enter into a marriage covenant. Remember, both fellowship and marriage are meant to be driven by the goal of getting to know each other more over time. If the couple is at fellowship before they reach marriage, then nothing changes in their ultimate goal; they just add the benefit of being able to add an area where they can grow in getting to know each other over time: the physical.

The longer answer, for those people who really want to step up their "getting to know each other" game a few notches, is for them to focus on their "no touch" game. I've seen this myself.

The couple spends the first year of their relationship without intimate physical touching of any kind. This allows them to focus on their fellowship (getting to know who you ARE) and removes the primary source of energy people get from their future spouse (physical touch) that they can often confuse with love. Unfortunately for most couples, when intimate physical touching begins, that is when they stop getting to know each other and stop growing in fellowship.

It can be difficult, but it all comes down to the true goal of the couple. Are they really trying to get to know each other in who they ARE, or are they looking for some short-term and physical benefit from their partner? It's not wrong to want to physically touch your partner in an intimate way, but it can be a major distraction from the mental, emotional, and spiritual intimacy that ought to be the driver in getting to know each other.

My friend Jonathan likes to coach couples in this area by telling them, "Whatever it is you want to touch on your partner with your hands, put those feelings into words that you share with them." If you can do that, it's a sign that you're intimately touching the person for their benefit, not yours. That brings up two more important and non-contradictory definitions:

Sex: intimately touching another person for *their* benefit.
Groping: intimately touching someone for your *own* benefit.

Which of these, sex or groping, is key to a profitable marriage?

CHAPTER 8

Divorce

WHAT IS BIBLICAL divorce? In *Modeling God*, I wrote the following with regard to when it is right to end marriage covenants: *We will look at this in detail in a subsequent book.* It's time to look at this in detail!

How was the marriage covenant broken according to the Bible?
How was the marriage covenant ended according to the Bible?

The words used for divorce, divorced, and divorcement in the Old Testament are:
    H3748 keriythuwth — cutting of (matrimonial bond) from
    #3772 karath — cut (off, down, or asunder)
    H1644 garash — drive out

The word used in the New Testament is:
    G647 apostasion — something separative, specifically writing of divorcement from #868 aphistemi — to remove, instigate to revolt

Divorce is a definite ending of covenant and community between a husband and wife. It is not *lukewarm*. Divorce is accomplished by a statement of each one's will that they are no longer in agreement with the covenant. So, one of the real world circumstances that would prevent a divorce is one of the parties enjoying controlling the other party, which makes it possible to see the other party as the enabling party. Consequently, the controlling party won't state their will to end the covenant, preventing the divorce for the enabling party.

We already saw that marriage in the Bible was accomplished without the government. When did the government and church get involved in divorce?

While the church and government only got involved in marriage 300-500 years ago, the church and government got involved in divorce more than a thousand years before they got involved in marriage.

The Christian emperors Constantine and Theodosius restricted divorce to matters of "grave cause," but this was relaxed in the sixth century by Justinian. Four to five hundred years later, the church considered marriage a sacrament initiated by people but regulated by the church. The church allowed annulment and separation, but complete termination of the relationship was generally not allowed. The government had no role in divorce.

When the government did get involved in marriage, those governments had no precedent for divorce, so they took their cue from the church. The government considered divorce to be against the public interest, so the requirements for divorce returned back to "grave cause." Divorce could only be granted if one and only one of the parties was guilty. Divorce was denied if both parties were guilty!

Divorce was also denied if the courts believed that the couple had worked together to manufacture grounds for divorce or even if the court thought they even attempted to create grounds for a divorce!

There are two possible wills God has for unprofitable marriage. God either wants people in unprofitable marriages to set an example to following generations by:

- enabling the controlling party until all the profitability is completely removed from the marriage
- ending the covenant once it is proven the marriage cannot attain profitability after the controlling party has rejected the covenant ending process. In this latter case, the enabler is seen as a victim of abuse.

The history of divorce shows that the church and the government were focused on limiting the ending of the marriage covenant until all the profitability was completely gone. In fact, it seems that ending the marriage covenant was prohibited even when both parties were completely unprofitable. What does God's word say about divorce?

Biblical Divorce

In the rest of this chapter, I will list all of the Old Testament scriptures using the word divorce (and its derivatives). There actually aren't many, so it is possible to definitively understand God's perspective on divorce.

Let's begin with the law listing the explanation for *how/why* to divorce:

> *When a man hath taken a wife, and married her, and it come to pass that she find no favour in his eyes, because he hath found some uncleanness in her: then let him write her a bill of divorcement, and give it in her hand, and send her out of his house. And when she is departed out of his house, she may go and be another man's wife. And if the latter husband hate her,*

and write her a bill of divorcement, and giveth it in her hand, and sendeth her out of his house; or if the latter husband die, which took her to be his wife; Her former husband, which sent her away, may not take her again to be his wife, after that she is defiled; for that is abomination before the Lord: and thou shalt not cause the land to sin, which the Lord thy God giveth thee for an inheritance. When a man hath taken a new wife, he shall not go out to war, neither shall he be charged with any business: but he shall be free at home one year, and shall cheer up his wife which he hath taken. (Deuteronomy 24:1-5)

A bill of divorcement was written and the woman was allowed to become another man's wife. She remained in community with the whole (church). She was no longer in the community as it related to marriage.

The reason (*why*) was: no favor in the husband's eyes because of some uncleanness. This was very general. Notice, no one was forcing these people to stay together until they were massively unprofitable. Also, remarriage of a divorced person was completely acceptable and not seen as a sin! Where did we get the view that divorce is a sin or something to be looked down on? Is this another reason it was against the law to translate the Bible into English?

However, once a man divorced a woman and she got divorced or widowed from another man, the first man couldn't remarry her. That was a sin according to the law. It seems this policy would discourage someone from jumping to the *cold* option (divorce) without first trying the *hot* option (make the marriage profitable).

In Leviticus, the law concerning priests (Levites) marrying was as follows:

> *And he shall take a wife in her virginity. A widow, or a divorced woman, or profane, or an harlot, these shall he not take: but he shall take a virgin of his own people to wife.* (Leviticus 21:13-14)

Notice, this law treated a divorced woman like a widow and every other woman who was not a virgin. The woman was still in community and, in some ways, equal to a widow, while the profane and harlots were not supposed to be a part of the community.

In the next chapter of Leviticus, a twist was added:

> *If the priest's daughter also be married unto a stranger, she may not eat of an offering of the holy things. But if the priest's daughter be a widow, or divorced, and have no child, and is returned unto her father's house, as in her youth, she shall eat of her father's meat: but there shall no stranger eat thereof.* (Leviticus 22:12-13)

The priest's daughter was not able to eat from the holy things if she was married to a stranger (unbeliever). However, she could eat of the holy things if she returned to her father's house even after she was divorced. The phrase "as in her youth" meant when she was a virgin.

Notice, being in community with an unbeliever was more limiting than being divorced! This passage was actually prioritizing the meaning of life. Community with the whole (church) was limited if the individual had community (marriage) with a person outside the whole.

However, getting divorced doesn't remove her from the greater community (church), it actually restored her fully to community! Today, do we restore people to community after they are divorced?

The same mentality occurred in the Book of Numbers when it came to a woman making a vow:

> *If a woman also vow a vow unto the Lord, and bind herself by a bond, being in her father's house in her youth; And her father hear her vow, and her bond wherewith she hath bound her soul, and her father shall hold his peace at her; then all her vows shall stand, and every bond wherewith she hath bound her soul shall stand. But if her father disallow her in the day that he heareth; not any of her vows, or of her bonds wherewith she hath bound her soul, shall stand: and the Lord shall forgive her, because her father disallowed her. And if she had at all an husband, when she vowed, or uttered ought out of her lips, wherewith she bound her soul; And her husband heard it, and held his peace at her in the day that he heard it: then her vows shall stand, and her bonds wherewith she bound her soul shall stand. But if her husband disallowed her on the day that he heard it; then he shall make her vow which she vowed, and that which she uttered with her lips, wherewith she bound her soul, of none effect: and the Lord shall forgive her. But every vow of a widow, and of her that is divorced, wherewith they have bound their souls, shall stand against her.* (Numbers 30:3-9)

When she was a virgin ("in her youth"), her father had final say. When she was married, her husband (covenant partner) had final say. However, divorced women and widows were treated the same and spoke for themselves because they had no covenant partners.

So far, these passages concerning the law have shown there was a process for divorce and remarriage without it being declared a sin. When did people create the tradition that divorce needed to be difficult and treated as a sin? When did our need to keep up appearances turn marriage from a benefit to a prison sentence? What does this look like from God's perspective?

God's Perspective

When the couple makes a vow in church, God considers them married (betrothed) if they haven't had intercourse. If they have had intercourse before the vows, then God sees the covenantal intercourse as marriage and the ceremony as tradition.

Actually, God looks at the vows at the traditional ceremony as an opportunity for the enemy to attack the couple. Think about the vows recited at a traditional wedding ceremony.

The reality is that both people break their vows within months of the wedding. For example, when a person doesn't love, honor, or cherish their spouse (even once), they have broken the vows and the covenant. This actually allows the enemy a foothold to attack the marriage. The other spouse has the right to end the covenant once the covenant is broken.

The other situation that commonly occurs applies to the couple that has consensual intercourse for the first time. They may have just met at a party or a bar and now God considers them married. Having intercourse in this situation is not a sin in and of itself, it is marriage.

If the next night (or after) the groom has intercourse with another woman, he has committed adultery. That is a sin. Until he officially states his will that he is no longer in covenant (agreement) with his spouse (the first woman) to his spouse, every act of intercourse with someone else is adultery.

Think back to the traditional view of marriage and divorce. Would it make sense to force everyone to find the first person they had intercourse with and compel them to get legally married? God considered them married. Why don't we? Because of our tradition and culture and appearance, which is fornication!

However, from the view of the levels of sharing, intercourse instantly puts a person on the marriage covenant level. This doesn't necessarily mean they are married in God's eyes. For example, adultery isn't marriage, but it still places the adulterers at the marriage covenant level of sharing.

In the above example, the last person he had intercourse with would actually be on the marriage covenant level as it relates to sharing. The previous occupant of the marriage covenant level would return to their previous level, most likely connection or relationship. However, God would still see the original covenant partners as married until one of them ended the covenant.

Making Divorce Easy?

There are those who will say that if we make divorce easy, then people will just constantly marry and divorce, they will be able to have sex with whomever they want as long as they declare themselves divorced immediately afterwards.

First, this is already happening without the benefit of understanding what God's word said about divorce. Second, we saw that one way to ensure profitability at the marriage covenant level would involve being at fellowship for a year without touching. If people want to spend a year at fellowship without touching, then have intercourse, and then immediately divorce, that would be completely their choice.

When people accuse me of making divorce easy, they are trying to make man's doctrine into God's doctrine by barring people from doing anything apart from man's doctrine. This is a resolve approach. Making divorce difficult won't result in God's doctrine. We saw it will just empower the controller and trap the enabler. Prohibiting divorce actually encourages abuse! Anything less than dissolve is going to create three more problems.

Again, what is the objective of these people? Is it to help people achieve God's doctrine or reinforce man's doctrine?

We've seen Deuteronomy 24 gives provisions for divorce by issuing a bill of divorcement. What is the purpose of the bill of divorcement? Take a second to come up with your answer before continuing.

The bill of divorcement serves the same purpose as the bloodstained sheet! The sheet of paper definitively stated she wasn't in covenant with someone else. That is a dissolve approach, but most people don't see it because they don't realize what is really going on. Look at Deuteronomy 22 again (bloodstained sheet passage) and see what was really going on:

> *If any man take a wife, and go in unto her, and hate her, And give occasions of speech against her, and bring up an evil name upon her, and say, I took this woman, and when I came to her, I found her not a maid: Then shall the father of the damsel, and her mother, take and bring forth the tokens of the damsel's virginity unto the elders of the city in the gate: And the damsel's father shall say unto the elders, I gave my daughter unto this man to wife, and he hateth her; And, lo, he hath given occasions of speech against her, saying, I found not thy daughter a maid; and yet these are the tokens of my daughter's virginity. And they shall spread the cloth before the elders of the city.* (Deuteronomy 22:13-17)

Remember, the law said he could divorce her and give her a bill of divorcement. That was not what was happening here. His intent was to remove her from society. This was serious. Read the rest of the passage:

> *And the elders of that city shall take that man and chastise him; And they shall amerce him in an hundred shekels of silver, and give them unto the father of the damsel, because he hath brought up an evil name upon a virgin of Israel: and she shall be his wife; he may not put her away all his days. But if this thing be true, and the tokens of virginity be not found for the damsel: Then they shall bring out the damsel to the door of her father's house, and the men of her city shall stone her with stones that she die: because she hath wrought folly in Israel, to play the whore in her father's house: so shalt thou put evil away from among you.* (Deuteronomy 22:18-21)

If he was right, she was killed! There was no mention of divorce. He was trying to get her killed!

If he was wrong, he can't ever (*all his days*) "put her away." What did that mean?

What is putting away?

Dr. Joel Swokowski's Commentary

Laws of the Land?

In the Marriage chapter, I wrote the following in the commentary section:

One of the opposing arguments against the "covenantal sexual intercourse is marriage" that we've come across is "we're supposed to follow the laws of the land."

First, although this concept is in the scriptures (Romans 13:1-2), is it meant for us to follow the laws of the land at the expense of God's commands?

Generally speaking, laws tell you what you cannot do, not everything you ought to do. It's not against the law of the land to embrace and believe God's definition for marriage. It may result in the government not recognizing the marriage, but that doesn't mean the husband and wife aren't married.

Thinking about divorce, it would only be fair if these same people who appealed to following the laws of the land also allowed for divorce, since divorce is allowed by the laws of our land. However, this is a great example of when "Christians" want to approach topics in a piecemeal fashion, allowing for differing standards for how an issue is approached. Why is it acceptable to use a passage of scripture for dealing with marriage but ignore the scriptures when dealing with divorce?

Evangelist Tool?

I've heard, and read, from Christian leaders and books on marriage that divorce is a sin and should never happen; that Christian marriages, when struggling, should just "stick it out" and appear to have a happy marriage for the purpose of presenting a good Christian life. This "sticking it out" was encouraged to be used as an evangelist tool.

What message is this getting across?

- God wants unprofitable marriage?
- God is forcing you to be in an abusive relationship?

- Being a Christian and in a Christian marriage means having no way out, even if it's bad?

Parents are modeling marriage to their kids. When they stay in a bad marriage for the kids, they are promoting and normalizing bad marriages. It gets worse when we force (or attempt to force) people to stay in bad marriages for the sake of a man-made tradition. Staying in a bad marriage or forcing others (including young men and women) to stay in their bad marriages is perpetuating a man-made doctrine that is hindering the will of God.

Parental Concern

A common concern about parents in our day and age is premarital sex. We've already seen that this concept is self-contradictory in that consensual sexual intercourse *is* marriage. However, due to the world's (and church's) ignorance of God's view of marriage, there is a huge effort to prevent "premarital sex."

When I look at some of the most popular methods being used, I honestly wish I was more surprised at what is being taught to teens. Is it any wonder that "premarital sex" is on the rise when the following are the main techniques used by the church?

- Stay in church
- Read your Bible
- Don't wear promiscuous clothing
- Don't go out alone with your date
- Don't go places where you could be tempted
- Stay away from sexually evocative material
- Accountability by parents
- Be committed to sexual purity

Notice, all eight of these techniques are either telling a person what *not* to do (and you can't do a don't), or they are too abstract and do not give you a method to (*how*) or reason (*why*) that actually helps the teens. In fact, most of these would simply increase the likelihood of "premarital sex."

The most popular techniques are to scare the teens by telling them they are going to get a disease or a pregnancy, which is treated like a disease. Neither of these thoughts are strong enough to deter the majority of teens from "premarital sex."

We've found the best approach is in teaching teens that consensual sexual intercourse is marriage. I've seen it myself, teaching a teenage boy that he would be married to the girl he plans to have sex with, and that he'd be responsible to lead her as a man — it takes the energy right out of him. Teaching this to a teenage girl usually results in her simply being settled and putting words to her feelings (mainly because females are naturally more mature than males, they "get it" quicker).

Once again, God's doctrine results in a dissolve approach.

CHAPTER 9

Putting Away

WHAT IS THE doctrine of putting away?

Is putting away the same as divorce?

The word used for putting away in the Old Testament was:
    H7971 shalach — to send away, for, or out
The word used for putting away in the New Testament was:
    G630 apoluo — to free fully, relieve from #575 apo — off, away
    and #3089 luo — loosen

Putting away: shalach and apoluo are passive and without conflict

Divorce: keriythuwth, garish, and apostasion are aggressive and violent

Putting away is not the same as divorce. These are two completely different actions that have two different causes and objectives.

Putting away is just to walk away from the person and move on to the next person. This is the definition of *lukewarm*. There was nothing definite about this. There was no paper trail; people didn't know if these women were married or not. This practice wrecked society.

However, putting away is the perfect process for removing foreigners and unbelievers from society! These women would never be able to marry in society and bear kids to that society. Putting away was used to remove unbelievers from the community. Notice this involved wives and children as seen in Ezra chapters 9 and 10. Let's cover the most crucial putting away verses in the Old Testament.

Remember, fornication means to commit idolatry. Idolatry is image worship. Fornicators are unbelievers. Keep this in mind as you read the following Bible passages.

The rest of the Old Testament divorce scriptures contained references to putting away. Let's begin with the most important passage and its implications.

> *Thus saith the Lord, Where is the bill of your mother's divorcement, whom I have put away? or which of my creditors is it to whom I have sold you? Behold, for your iniquities have ye sold yourselves, and for your transgressions is your mother put away.* (Isaiah 50:1)

God first put away Israel because of her transgressions. However, God took Israel back. He did not end the covenant at this point. Divorcing would be the definitive ending of the covenant.

We saw that God did end the covenant in Jeremiah 15, so God did put away and divorce Israel. However, how did it get to that point? Let's look at what happened before Jeremiah 15.

Here was a section from Jeremiah 3.

> *They say, If a man put away his wife, and she go from him, and become another man's, shall he return unto her again? shall not that land be greatly polluted? but thou hast played the harlot with many lovers; yet return again to me, saith the Lord.* (Jeremiah 3:1)

God was establishing that He was doing something different than man. Man puts away his wife and doesn't take her back. God was saying that even though He put away His "wife," He was inviting her back. God was offering to continue the covenant!

> *Lift up thine eyes unto the high places, and see where thou hast not been lien with. In the ways hast thou sat for them, as the Arabian in the wilderness; and thou hast polluted the land with thy whoredoms and with thy wickedness. Therefore the showers have been withholden, and there hath been no latter rain; and thou hadst a whore's forehead, thou refusedst to be ashamed. Wilt thou not from this time cry unto me, My father, thou art the guide of my youth? Will he reserve his anger for ever? will he keep it to the end? Behold, thou hast spoken and done evil things as thou couldest.* (Jeremiah 3:2-5)

God was stating that there were consequences to ending the covenant because of justice. God was trying to motivate His covenant partner not to stray anymore.

> *The Lord said also unto me in the days of Josiah the king, Hast thou seen that which backsliding Israel hath done? she is gone up upon every high mountain and under every green tree, and there hath played the harlot. And I said after she had done all*

> these things, Turn thou unto me. But she returned not. And her treacherous sister Judah saw it. And I saw, when for all the causes whereby backsliding Israel committed adultery I had put her away, and given her a bill of divorce; yet her treacherous sister Judah feared not, but went and played the harlot also. (Jeremiah 3:6-8)

Now, God was stating that Israel chose to continue to break the covenant, so God gave her a bill of divorce. He ended the marriage covenant. However, God's focus was on Judah.

> And it came to pass through the lightness of her whoredom, that she defiled the land, and committed adultery with stones and with stocks. And yet for all this her treacherous sister Judah hath not turned unto me with her whole heart, but feignedly, saith the Lord. And the Lord said unto me, The backsliding Israel hath justified herself more than treacherous Judah. (Jeremiah 3:9-11)

God stated that Judah pretended to return but didn't. So Judah was guiltier because she knew better (saw what happened to Israel) and faked her intentions. Take a second and look at the implications.

God divorced Israel, but was willing to take her back. He divorced Israel for Israel's benefit so that Israel would objectively see what would happen. This divorce didn't prevent God from remaking the covenant.

When tradition prohibits divorce, they are prohibiting a tool for growth. God used divorce as a tool to repair the marriage! Traditional leaders prevent us from using a God-given tool to repair marriages! Why? Because of appearance? Isn't that fornication?

In fact, prohibiting divorce guarantees *lukewarm* for both parties. The party that is in the wrong misses out on an objective way to find out they are wrong and the profitability of the innocent party is hindered. For example, how can the innocent party get their prayers answered if they aren't in agreement with the covenant partner (I Peter 3:7)?

Today, some people use this fact to force the wronged party (abused) into agreeing with the unprofitable member (controlling abuser) of the covenant. I have seen that the statement of wills from both individuals that they are no longer in agreement actually causes the innocent party to get their prayers answered! This is Biblical divorce! Even though the government and the church still see the couple as "married," the Bible said they are divorced. This is a God-given tool to repair marriages! How does this look in our marriages?

Repairing Marriage

When the marriage has an overwhelming number of unprofitable aspects, so many they outnumber the profitable aspects, and the ability to be profitable is not possible, this marriage ought to be ended. This would be Biblical divorce. The couple declares they are no longer in agreement with their covenant, and both then turn to God and pray for correction.

God would show both spouses where they have been wrong. If both spouses are doing the first command, then their doing the second command would involve repenting; giving a value to the other that attempts to make up for the value they owe due to the damaging things they have said and done.

It is only at this point that the spouses can forgive each other! Remember, Jesus said in Luke 17:3 that if the believer repents, then you forgive. Forgiving before repentance actually brings judgment on the offending party.

Notice, there are pastors and "Christian" authors telling married couples to "hit the reset button." This is a resolve approach. What they mean is that both people should just pretend that nothing bad has happened. They should forgive without repentance/repairing! This creates three more problems. This actually causes justice to bring judgment on the counselors! This also enables the abusive controllers! This leads to not sharing!

True repair would cause the couple to want to discuss the situation after the fact. When people truly repair, they both want to discuss the circumstances years later because they both gained. Remember, repair is "ultra sharing." Repair is an exchange (right *what*) that takes abuse (unprofitable sharing) and makes it profitable through sharing (right *how/why*). Sharing is a right *what* (exchanging information) with a right *how/why* (for the benefit of another/for understanding). Ideally, this would be mutually beneficial.

Tina committed adultery against Stan, her husband. Not only did Tina confess and repent to Stan, she has been willing to share her story with others so they don't make the same mistakes she made. Unfortunately, the initial, private counsel they received was for Stan to "hit the reset button" and just take Tina back in order to keep up the appearance of a good, Christian marriage because Stan is a leader in the church.

When people "hit the reset button," they don't want to discuss the circumstances ever again. That is the point of the "reset button." This is deception: a right *what* (pursuing growth) with a wrong *how/why* (ignoring facts in order to get a short-term appearance based benefit).

The real deception in this "reset button" technique is that the counselor will intentionally focus the couple on the profitable aspects of the past. The thought is that the couple can retain the profitable aspects and forget the unprofitable aspects. This is a right *what* with a wrong *how*; however, the *how* sounds right. The problem is that the unprofitable memories are tied to the profitable memories.

The only way for this couple to progress is to let go of all the past aspects, both profitable and unprofitable. This is a dissolve approach. The couple ought to truly treat each other like new creations. After the couple divorces (makes a clean break with the past), repents (repairs the past hurts), and forgives, they ought to begin the process as two completely new people forming completely new memories.

I have seen this God-given tool to repair marriages work, but remember, it takes both parties treating each other as new creations. This is the true way to "hit the reset button"! They haven't forgotten the old creations. In fact, I have told people to refer to their former selves with different names. They have begun something new and kept the lessons from the old. Let's get back to how God dealt with Judah and Israel.

God's Approach

Judah stayed married but not with her whole heart. So, Israel was divorced and Judah was married, yet God said Israel has justified herself more than Judah! How was this possible? This was because Israel was *cold* and Judah was *lukewarm*. People who think that remaining married for appearance is better than getting divorced are not in agreement with God's word. They are actually teaching something else in place of God's word and will. Let's continue with Jeremiah 3.

> *Go and proclaim these words toward the north, and say, Return, thou backsliding Israel, saith the Lord; and I will not cause mine anger to fall upon you: for I am merciful, saith the Lord, and I will not keep anger for ever. Only acknowledge thine iniquity, that thou hast transgressed against the Lord thy God, and hast scattered thy ways to the strangers under every green tree, and ye have not obeyed my voice, saith the Lord. Turn, O backsliding children, saith the Lord; for I am married unto you: and I will take you one of a city, and two of a family, and I will bring you to Zion:* (Jeremiah 3:12-14)

God ended the covenant so that Israel would realize the effects. Now God was saying He was willing to take her back if she confessed and repented. The word for married in this passage (H1166 baal) actually meant master of the covenant. God was in control of the covenant because Israel had backslid. He could choose to continue it or end it.

Israel has gone through fear of loss. Now God wants to show Israel what they will gain.

> *And I will give you pastors according to mine heart, which shall feed you with knowledge and understanding.* (Jeremiah 3:15)

God will give pastors that will feed people with knowledge and understanding. This was what would keep Israel profitable. Do we have pastors with knowledge and understanding? Do we have pastors who know the word of God and do it? Do we have pastors who understand the word of God? Pastors who can explain the *why*?

Next, God jumps to the everlasting covenant. God was basically saying His will was going to eventually happen, but it was up to Israel as to when. If not now, then one day in the future.

> *And it shall come to pass, when ye be multiplied and increased in the land, in those days, saith the Lord, they shall say no more, The ark of the covenant of the Lord: neither shall it come to mind: neither shall they remember it; neither shall they visit it; neither shall that be done any more. At that time they shall call Jerusalem the throne of the Lord; and all the nations shall be gathered unto it, to the name of the Lord, to Jerusalem: neither shall they walk any more after the imagination of their evil heart. In those days the house of Judah shall walk with the house of Israel, and they shall come together out of the land of the north to the land that I have given for an inheritance unto your fathers. But I said, How shall I put thee among the children, and give thee a pleasant land, a goodly heritage of the hosts of nations? and I said, Thou shalt call me, My father; and shalt not turn away from me. (Jeremiah 3:16-19)*

God had shown the ultimate benefit. Now, He's going to conclude with their current state.

> *Surely as a wife treacherously departeth from her husband, so have ye dealt treacherously with me, O house of Israel, saith the Lord. A voice was heard upon the high places, weeping and supplications of the children of Israel: for they have perverted their way, and they have forgotten the Lord their God. Return, ye backsliding children, and I will heal your backslidings. Behold, we come unto thee; for thou art the Lord our God. Truly in vain is salvation hoped for from the hills, and from the multitude of mountains: truly in the Lord our God is the salvation of Israel. For shame hath devoured the labour of our fathers from our youth; their flocks and their herds, their sons and their daughters. We lie down in our shame, and our confusion covereth us: for we have sinned against the LORD our God, we and our fathers,*

from our youth even unto this day, and have not obeyed the voice of the Lord our God. (Jeremiah 3:20-25)

We know the choice that was eventually made. God became weary with their repenting and ended the covenant (Jeremiah 15:6).

Summary

Before we went to God's word, we stated the following:

Basically, there are two possible wills for God to have with unprofitable marriages. God either wants people in unprofitable marriages to set an example to following generations by:

- enabling controllers until all the profitability is completely removed from the marriage
- ending the covenant once it is proven the marriage cannot attain profitability after the controller (abuser) has rejected the covenant ending process

Ultimately, the answer to God's will is found in God's word.

What did the Bible say about ending marriage?

What did the Bible say about breaking and ending covenants?

What did the Bible say about breaking and ending community?

Now we see that the Old Testament picked the second option.

The Bible clearly stated that divorce, in and of itself, was permitted and was not sin. It is our man-made tradition that has defined marriage and divorce in terms of the state and then projected our doctrine onto God.

Since man is unable to explain the purpose of marriage according to God (*why*) and *how* to intentionally achieve a profitable community, his solution is to reinforce man's tradition, which actually prevents our ability to achieve God's will.

Divorce, according to the verses we have seen from the Bible, is a definite and objective ending of the marriage covenant so that both parties can remain in community with the greater community of believers.

Putting away, according to the Bible, is the ending of the marriage covenant because one of the people is not a believer and the objective is to remove the member from the greater community of believers, which is the church.

In a sense, the doctrine of marriage serves the doctrine of church with both being joined in the doctrine of the meaning of life.

Let's look in more detail at divorce and putting away.

Dr. Joel Swokowski's Commentary

Early in this chapter, Lenhart stated the following:

Putting away is not the same as divorce. These are two completely different actions that have two different causes and objectives.

Here are the rest of the Old Testament verses to flesh out this doctrine of putting away.

In Ezra, the unbelievers (fornicators) were put away.

> *Now when these things were done, the princes came to me, saying, The people of Israel, and the priests, and the Levites, have not separated themselves from the people of the lands, doing according to their abominations, even of the Canaanites, the Hittites, the Perizzites, the Jebusites, the Ammonites, the Moabites, the Egyptians, and the Amorites. For they have taken of their daughters for themselves, and for their sons: so that the holy seed have mingled themselves with the people of those lands: yea, the hand of the princes and rulers hath been chief in this trespass. (Ezra 9:1-2)*

> *Which thou hast commanded by thy servants the prophets, saying, The land, unto which ye go to possess it, is an unclean land with the filthiness of the people of the lands, with their abominations, which have filled it from one end to another with their uncleanness. Now therefore give not your daughters unto their sons, neither take their daughters unto your sons, nor seek their peace or their wealth for ever: that ye may be strong, and eat the good of the land, and leave it for an inheritance to your children for ever. (Ezra 9:11-12)*

> *Now therefore let us make a covenant with our God to put away all the wives, and such as are born of them, according to the counsel of my lord, and of those that tremble at the commandment of our God; and let it be done according to the law. (Ezra 10:3)*

Putting away was used to remove unbelievers from the community. Notice, this involved wives and children!

Finally, we are back to Abraham. Remember, Abram took Hagar as wife.

> *Now Sarai Abram's wife bare him no children: and she had an handmaid, an Egyptian, whose name was Hagar. And Sarai said unto Abram, Behold now, the Lord hath restrained me from bearing: I pray thee, go in unto my maid; it may be that I may obtain children by her. And Abram hearkened to the voice of Sarai. And Sarai Abram's wife took Hagar her maid the Egyptian, after Abram had dwelt ten years in the land of Canaan, and gave her to her husband Abram to be his wife. And he went in unto Hagar, and she conceived: and when she saw that she had conceived, her mistress was despised in her eyes. And Sarai said unto Abram, My wrong be upon thee: I have given my maid into thy bosom; and when she saw that she had conceived, I was despised in her eyes: the Lord judge between me and thee. But Abram said unto Sarai, Behold, thy maid is in thy hand; do to her as it pleaseth thee. And when Sarai dealt hardly with her, she fled from her face.* (Genesis 16:1-6)

Did Abram divorce Hagar?

Did Abram put away Hagar?

Hagar came back and stayed with Abram and Sarai for over ten years. Then God struck a new covenant, and we now have Abraham and Sarah. Sarah gave birth to Isaac, and then Hagar was sent away again.

> *And Sarah saw the son of Hagar the Egyptian, which she had born unto Abraham, mocking. Wherefore she said unto Abraham,*

Cast out this bondwoman and her son: for the son of this bondwoman shall not be heir with my son, even with Isaac. And the thing was very grievous in Abraham's sight because of his son. And God said unto Abraham, Let it not be grievous in thy sight because of the lad, and because of thy bondwoman; in all that Sarah hath said unto thee, hearken unto her voice; for in Isaac shall thy seed be called. And also of the son of the bondwoman will I make a nation, because he is thy seed. And Abraham rose up early in the morning, and took bread, and a bottle of water, and gave it unto Hagar, putting it on her shoulder, and the child, and sent her away: and she departed, and wandered in the wilderness of Beersheba. (Genesis 21:9-14)

Did Abraham divorce Hagar?

Did Abraham put away Hagar?

CHAPTER 10

Divorce vs. Putting Away

WITH MARRIAGE MAKING up half of God's plan and will, it would seem that confusing divorce with putting away would have dire consequences. For example, the most abused divorce scripture is in the Book of Malachi. Most traditional Christians quote Malachi 2:16 as "God hates divorce" and imply divorce is a sin, despite what we have learned so far.

Part of the reason for the confusion is that different versions of the Bible state different things. We have been using the King James Version in this edition, which we used in the first edition (*Modeling God*). The American Standard Version is very similar to the King James Version, especially relative to the use of putting away. Both of these versions of the Bible are more than 100 years old, which means no one owns them (public domain).

The most popular versions of the Bible other than these two are less than 100 years old, so someone owns them. Worse, these newer versions of the Bible seem to consistently confuse divorce and putting away. Here was Malachi 2:16 in the American Standard Version:

> *For I hate putting away, saith Jehovah, the God of Israel, and him that covereth his garment with violence, saith Jehovah of hosts: therefore take heed to your spirit, that ye deal not treacherously.*

While the King James Version also used putting away, the newer versions all used divorce in place of putting away. Here is a quick review of the definitions of each from the previous two chapters:

Putting away: shalach and apoluo are passive and without conflict

Divorce: keriythuwth, garish, and apostasion are aggressive and violent

First of all, given everything you have read, how is God "hating" divorce even possible? If God was completely against divorce, wouldn't that contradict the divorce scriptures we've already looked at? How could God be against divorce, yet He divorced Israel? If someone used this verse to say divorce is a sin, wouldn't they be stating that God sinned in divorcing Israel?

Second, the Hebrew word that was used actually meant putting away, not divorce. Why do people hold to a wrong interpretation of this one verse even though it makes several other verses wrong? Here was the whole passage in the King James Version:

> *Judah hath dealt treacherously, and an abomination is committed in Israel and in Jerusalem; for Judah hath profaned the holiness of the Lord which he loved, and hath married the daughter of a strange god. The Lord will cut off the man that doeth this, the master and the scholar, out of the tabernacles of Jacob, and him that offereth an offering unto the Lord of hosts. And this have ye done again, covering the altar of the Lord with tears, with weeping, and with crying out, insomuch*

that he regardeth not the offering any more, or receiveth it with good will at your hand. Yet ye say, Wherefore? Because the Lord hath been witness between thee and the wife of thy youth, against whom thou hast dealt treacherously: yet is she thy companion, and the wife of thy covenant. And did not he make one? Yet had he the residue of the spirit. And wherefore one? That he might seek a godly seed. Therefore take heed to your spirit, and let none deal treacherously against the wife of his youth. For the Lord, the God of Israel, saith that he hateth putting away: for one covereth violence with his garment, saith the Lord of hosts: therefore take heed to your spirit, that ye deal not treacherously. (Malachi 2:11-16)

"Wife of thy youth" meant the wife of their virginity; the first person they had intercourse with. God even called it the "wife of thy covenant." God loves less (hates) that these people have put away the wife of their youth by moving on to other women. This passage (with the rest of the Old Testament) actually implied God would prefer divorce over putting away! God loves putting away less than the alternatives.

In reality, God was upset because these priests were removing believing women from the community. Their putting away their wives prevented their wives from remarrying and the priests' new marriages were actually adultery because God still saw the priest as married to the wife of their covenant. Instead of understanding God's word and not hindering God's will, tradition takes a resolve approach by trying to prevent divorce by interpreting hate as completely against and putting away as divorce.

New Testament

In the New Testament, the first usage of putting away occurred in the first chapter (Matthew 1:18-19):

> ¹⁸ *Now the birth of Jesus Christ was on this wise: When as his mother Mary was espoused to Joseph, before they came together, she was found with child of the Holy Ghost.*
> ¹⁹ *Then Joseph her husband, being a just man, and not willing to make her a public example, was minded to put her away privily.*

There was a lot of doctrine in these two verses! First, let's look at verse 18. The word espoused is G3423 mnesteuo - "to give a souvenir."

Joseph had not had intercourse with Mary. They were betrothed because they exchanged tokens (souvenirs)! So Joseph was Mary's husband. However, the marriage hadn't been consummated; Joseph and Mary had not had intercourse. This interpretation is non-contradictory and meshed with the rest of the Bible. If you don't believe it, look at this passage from the law (which was what Joseph and Mary lived under):

> *And if a man entice a maid that is not betrothed, and lie with her, he shall surely endow her to be his wife.* (Exodus 22:16)

A maid (virgin) could be betrothed without having had sex. Mary was a maid that was betrothed. Now we can look at verse 19.

Joseph thought she wasn't a virgin (maid) because she was pregnant. Joseph believed Mary was married to someone else. Notice, Joseph was completely in the right to draw these conclusions. How did he know he was the *only* person who had to deal with the one exception (virgin birth)?!?!

The verse said Joseph was just for being minded to put her away. She appeared to be married to someone else. It would be wrong for Joseph to divorce her; to testify that she wasn't married to anyone. Joseph wouldn't do this because he was just. The Greek word for just was interpreted as right in about half of the versions of the Bible. We have seen in *Modeling God* that the choice for interpretation depended on whether the context implied if it was a qualitative (right) or quantitative (just) description.

The fact that he wanted to do it privily comes from not willing to make her a public example. Joseph's intent was to put her away because it was right and just to remove her from the community so someone else didn't marry an unbeliever; however, he had a choice to make. He could make a public example of her, or he could do it privately.

Remember the passage from Deuteronomy that dealt with a husband finding out his wife was not a virgin? He could have her stoned! That was a public example! Joseph was avoiding this public example by putting her away privately.

Again, Joseph showed he was a just man by not taking out his own justice. He was not interested in humiliating her so that he didn't look bad. He was going to bear the pain of looking bad when people realized they weren't married.

One more point, Joseph was also just for not having sex with her! Having sex with her would make him not a virgin and an adulterer. Joseph could have thought, "Who would know?" Joseph actually sounded like a person of high character; a right and just man.

I don't want to take away from what Mary went through; my point is that I don't think many people stop to consider that God's choosing

Mary had as much to do with Joseph as it did with Mary. There were several opportunities for Joseph to exert his will and wreck God's plan. God needed a just man as well as a righteous and pure virgin to bring about His plan.

The summary: Joseph was betrothed to Mary through an exchange of tokens; however, they had not consummated the marriage. When Joseph found out that Mary was pregnant, Joseph thought she was married to someone else. His only option at that point was putting away, not divorce. Every other explanation for these verses is contradictory.

Jesus and Pharisees

And the Pharisees came to him, and asked him, Is it lawful for a man to put away his wife? tempting him. And he answered and said unto them, What did Moses command you? And they said, Moses suffered to write a bill of divorcement, and to put her away. And Jesus answered and said unto them, For the hardness of your heart he wrote you this precept. But from the beginning of the creation God made them male and female. For this cause shall a man leave his father and mother, and cleave to his wife; And they twain shall be one flesh: so then they are no more twain, but one flesh. What therefore God hath joined together, let not man put asunder. And in the house his disciples asked him again of the same matter. And he saith unto them, Whosoever shall put away his wife, and marry another, committeth adultery against her. And if a woman shall put away her husband, and be married to another, she committeth adultery. (Mark 10:2-12)

There was something more going on here. There were four possible questions that could have been asked of Jesus:

1. When is it wrong to divorce?
2. When is it right to divorce?
3. When is it wrong to put away?
4. When is it right to put away?

In this passage, Jesus was asked question #4: When is it right to put away? The question was posed open-ended: Is it lawful for a man to put away his wife? It is like asking if there was *any* reason a man could put away his wife according to the law.

We've looked at all of the scriptures concerning divorce and putting away in the law. The cause for divorce was specifically stated in the law: no favor in the eyes of the husband due to some uncleanness in the wife, so there was a definite answer to questions #1 and #2. Jesus could also answer when it is wrong to put someone away, which was question #3.

Jesus was asked the only question He couldn't directly answer with scripture because the Bible didn't specifically state the cause(s) for putting away. We saw the answer was unbelief as an effect of fornication; however, that was not specifically stated in the law.

This was essentially a trick question! The Pharisees were trying to get Jesus to overextend Himself in His explanation of the law or state He couldn't answer the question, in order to create doubt in the people. The Pharisees wanted to prove Jesus was not the Son of God by stating He was wrong or He didn't know something that they knew.

There was no way to answer the question definitively, and the only way to give a definitive answer was to make something up. What would have been right and just for Jesus to say at that time? Remember, even though these are New Testament books, Jesus was still speaking during

Old Testament times; during the dispensation of the law. The law was still in effect and Jesus needed to fulfill the law.

Jesus answered the question with one short question: What did Moses command you? Again, He didn't answer the Pharisees' question directly, and this was not a sin.

The Pharisees answered with a statement: Moses said to write a bill of divorcement and put her away. Jesus responded by correcting them stating that the believers who were married (what God had joined together) ought not put away believing spouses (let not man put asunder). That was the same point of the Malachi passage! Notice, Jesus focused them on a question He could answer: #3. When is it wrong to put away?

Jesus' response was to speak about the ideal situation: if people chose to be married, God intended a man to cleave to his wife. Jesus went further and said that what God had *joined together, let not man put asunder*. Let's take a moment to truly understand what Jesus was saying because this takes some thought.

First of all, marriage is not a command. The law did not say everyone had to get married. We have seen that marriage is a God-given blessing that is an opportunity to generate spiritual value to bring about God's will. However, it is ultimately a choice. It is a choice to enter into marriage. It is a choice to exchange value with a spouse and God in a manner that generates spiritual value. It is a choice to use this spiritual value to bring about God's will.

So, Jesus was speaking about the ideal case when it came to all three choices. In fact, if you think about Jesus' answer more deeply, He was saying that there was a way for a man and woman to want to stay together. Would Jesus say, "God intended a man and a woman to stay together even though there was no way for these two people to have a marriage that continually grows in intensity and energy?"

Jesus was specifically talking about a marriage that God had joined together. Who would God join together? Believers! God told Israel to put away the unbelieving wives and children. So, the first point to realize about this passage was that Jesus was only speaking about marriage between believers!

Again, tradition tells us that God has joined every marriage together. Isn't this the same as making our will into God's will? Isn't this prideful? What did "joined together" mean?

The phrase "joined together" is G4801 suzeugnumi — "to yoke together"!

> Be ye not unequally yoked together with unbelievers: for what fellowship hath righteousness with unrighteousness? and what communion hath light with darkness? (II Corinthians 6:14)

Jesus was saying that the husband and wife were in their proper roles and they were equally yoked. They were believers! These two people were in fellowship. We know the only way they can be in fellowship with each other is if they were in fellowship with God. This was proof God had yoked them together. People who were not in fellowship with each other were not joined together by God. They were not both believers!

The second half of the verse said let no man put asunder what God had yoked together. What did "put asunder" mean?

> put asunder is G5562 choreo — from 5561; to be in (give) space G5561 chora — fem. of a der. of the base of 5490 through the idea of empty expanse

Jesus was telling the man not to put an empty (unprofitable) space between him and his wife. This was putting away! This is consistent with everything we've seen: Believing spouses shouldn't be put away. If they were ending the marriage covenant, they ought to divorce. Jesus could answer this question definitively.

Part of the confusion people experience with these passages occurs because of the progression of the discussion:

- The Pharisees' original question concerned putting away.
- The response from Jesus was distracting because it caused the Pharisees to bring up divorce.
- Jesus' conclusion went back to a putting away question He could answer definitively.

Newer versions of "the Bible" change the discussion to make the conversation appear more consistent. They tend to keep the conversation on divorce by changing the Pharisees' original question and Jesus' final statement, whether Jesus' final statement occurred in front of the Pharisees or in private with the disciples.

Every "Bible" is a worldview. When a person proves a point by using several versions of "the Bible," it is the same as someone proving a point by picking and choosing from several religions and saying it's truth. This is called "version shopping"; shopping for the version that proves your point instead of finding out the point God was proving. This is a justification of self. This is idolatry. This hinders God's will.

The non-contradictory word of God said putting away is not acceptable between believers and is only acceptable when fornication (covenanting with an image, which is unbelief) is committed. The contradictory "Bibles" change these verses to say divorce is only acceptable when adultery is committed. This is more than an embarrassment. It hinders the will of God.

Jesus' Identity

Once the encounter with the Pharisees was over, the next verse stated that His disciples asked Him again of the same matter in private. This makes sense why the disciples asked Him again, because He didn't answer the original question! There was no mention of the right *how/why* to put away (question #4). Notice also, that this answer was from both the husband's and wife's perspectives.

In private, Jesus answered question #3: When is it wrong to put away? Here is Mark 10:11-12.

> *And he saith unto them, Whosoever shall put away his wife, and marry another, committeth adultery against her.*

> *And if a woman shall put away her husband, and be married to another, she committeth adultery.*

Newer versions substitute divorce for putting away in both verses. Do you notice anything strange with these newer versions?

According to the King James and American Standard versions, a woman can put away her husband. According to the other versions, they are stating that Jesus told His disciples that a woman could divorce a man! There are so many things wrong with this I don't know where to begin.

Did the man need a bill of divorcement? Why? What would it look like for a woman to divorce a man? Was Jesus changing the law? Why didn't

His disciples question Him further on this? Why didn't they leave Him because Him stating a comment like this would have proven He's not the Son of God? Notice the other half of this passage.

According to the King James and American Standard versions, a man commits adultery if he marries after putting away his believing wife. This was the case we saw in Malachi. According to the other versions, Jesus stating a comment like this would have been Him telling His disciples that a man commits adultery if he marries after divorcing his wife!

The law said it was not a sin to marry another woman after divorcing, which means these other versions present a Jesus who said that doing the law resulted in sin! Worse, since God divorced Israel and married us, these versions have Jesus stating that God is an adulterer!

These "versions of the word of God" have Jesus not only contradicting the law but proving He doesn't know the law by stating that doing the law results in sin. How is it possible for Jesus not to know the word of God? How is it possible for this Jesus to be the Son of God? Why do we hold to doctrine from a presentation of a Jesus that isn't the Son of God?

When people use these "Bibles" to define divorce as "divorce due to adultery or unbelief ONLY" it is actually a confession that person doesn't believe in Jesus' divinity! When someone calls this "Biblical" divorce, they are stating their will as to what "Bible" they believe. If it is a "Bible" that said Jesus was wrong or misled people, then these people are stating their will they believe in an interpretation that would prove to us Jesus was not the Son of God.

When people reference the following versions as "the word of God" (NASB, NIV, NKJ, The Message, NLT, and ESV) and base their doctrines on them, they are confessing that they are Pharisees and Jesus is not the Son of God.

I've written to the publishers of these versions and asked them if they think it is right to publish a book that proves Jesus is not the Son of God. Three responded. None of them answered my question, blaming interpretation and not translation, while continuing to make a monetary profit on a book presenting man's doctrine and claiming it to be God's doctrine.

Summary

God created both marriage and church as a way for us to be profitable and generate the spiritual value needed to bring about God's will! We have seen the importance of following God's word: it is profitable and leads us to bring about God's will. We have seen profitability is measured by communication and intimacy.

The last five chapters have shown how man's tradition has been put in place of God's doctrine of marriage, divorce, and putting away.

How will you respond to people who are hindering God's will by defining marriage according to man's tradition?

It took us five chapters to answer the question posed in *Modeling God* as to when it is right to end a marriage covenant. Jesus was able to answer the Pharisees' question in less than a handful of sentences. It was vital that we covered the answer in depth as it relates to half of God's plan and will.

In fact, now we can see the previous representation of God flowing through an individual on the left half of the wall is marriage. God is flowing through two people so they exchange with each other in their uniqueness.

Let's look at the other half of the wall to see more of God's plan and the way God created for us to bring about His will.

Dr. Joel Swokowski's Commentary

A similar encounter happens with Jesus and the Pharisees as recorded by Matthew, but it is a different encounter based on the progression of the discussion.

> *The Pharisees also came unto him, tempting him, and saying unto him, Is it lawful for a man to put away his wife for every cause? And he answered and said unto them, Have ye not read, that he which made them at the beginning made them male and female, And said, For this cause shall a man leave father and mother, and shall cleave to his wife: and they twain shall be one flesh? Wherefore they are no more twain, but one flesh. What therefore God hath joined together, let not man put asunder. They say unto him, Why did Moses then command to give a writing of divorcement, and to put her away? He saith unto them, Moses because of the hardness of your hearts suffered you to put away your wives: but from the beginning it was not so. And I say unto you, Whosoever shall put away his wife, except it be for fornication, and shall marry another, committeth adultery: and whoso marrieth her which is put away doth commit adultery.* (Matthew 19:3-9)

In this passage, Jesus was asked #4: When is it right to put away? The Pharisees asked if it is lawful (right) for a man to put away his wife for *every* cause.

Jesus' first answer was a question focused on God's will and a statement that caused the Pharisees to ask a question focused on *why*. Then Jesus was able to make a statement that answered with both the right and wrong *how/why* to put away a spouse from the point of view of the male.

The Pharisees' response was to ask the *why* behind question #2 (When is it right to divorce?). The Pharisees asked Him *why* Moses commanded men to give a bill of divorcement and put away their wife. The Pharisees have now introduced the topic of divorce.

Take a moment and look at the strategy being used with relation to questions and statements. The Pharisees asked a question (which isn't a sin), in hopes of getting a wrong statement from Jesus. Jesus did not answer their question, but actually phrased God's will in the form of a question and ended with a factual statement relative to the law: Believers shouldn't put away their believing spouses. The Pharisees' response was to ask another question; however, it was different from their original question.

Again, the Pharisees were trying to trap Jesus. Jesus couldn't answer their first question with a statement without being wrong because the direct answer was not in the law. Jesus' response caused the Pharisees to ask a question about the law. Could Jesus now answer this question?

They wanted to know *why* Moses gave them the command to divorce and the command to put away if God didn't intend for it to happen. They were really asking a question about God's will: *Why* are we allowed to do something against God's will? The answer to that was that God didn't force His will on us. God's will doesn't always happen, and that is because of the hardness of our hearts.

Remember, the law did not command us to get married. It is our choice to marry. Although marriage is intended to be a blessing, it isn't guaranteed. If marriage isn't profitable, that is our choice. However, God wouldn't want a person to live in unprofitability, so God provided for divorce. There is a huge implication here that most people miss.

Actually, this passage proved divorce is a blessing! Look at it this way: An unprofitable marriage is like a sickness. However, God provides a cure: divorce. Does this make divorce a sin? Is medicine a sin or a blessing? We wouldn't need medicine if we weren't sick. However, our sickness doesn't make the medicine bad.

Let's take this to an extreme: Is Jesus' sacrifice on the cross bad? Is Jesus' death on the cross a sin? The only reason He did it is because we needed it because of our "sickness." Stating that divorce is "bad" or "a sin" is the same as stating Jesus' death on the cross is "bad" or "a sin." Be careful, Jesus was saying that God provided a solution (divorce) because we caused a problem (hardness of our hearts).

Again, Jesus turned the answer back on them and talked about God's will. (*Moses because of the hardness of your hearts suffered you to put away your wives: but from the beginning it was not so.*) The *why* has to do with man's will: people don't have soft hearts and take direction from God. Moses provided a way to deal with the fact that people have hard hearts. Jesus has now answered the first question (#4) the Pharisees asked without falling into their trap!

There are only two places in the Old Testament where divorce and putting away were mentioned together. Neither involved Moses telling people to divorce *and* put away.

The Pharisees were trying to imply that Moses was encouraging people to divorce and put away their wife; that Moses was making divorce easy! People who accuse Lenhart of this today are acting like the Pharisees!

Notice, Jesus' answer focused on the judgment of the males: current husband and future husband. Notice also, He focused the males on the doctrine of putting away and divorce by talking about intercourse.

Jesus made a final statement to the Pharisees concerning putting away, fornication, and adultery.

What explanations have you heard for this passage?

New Testament

Some people will say that everything we have covered relative to divorce and putting away was only true under the law, however we see the same issues with a passage from Paul as what we saw Jesus deal with in Mark 10:2-12 and Matthew 19:3-9.

> *And unto the married I command, yet not I, but the Lord, Let not the wife depart from her husband: But and if she depart, let her remain unmarried or be reconciled to her husband: and let not the husband put away his wife.* (I Corinthians 7:10-11)

Paul said this command came from Jesus, which was the same thing Jesus said before his resurrection in Matthew 10:11,12. A believing wife and a believing husband shouldn't separate, however, if the believing wife walks away, encourage the husband not to put her away. They could divorce.

Notice also, Paul is writing this letter to the church, not the husband and wife. This becomes important in the next two verses:

> *But to the rest speak I, not the Lord: If any brother hath a wife that believeth not, and she be pleased to dwell with him, let him not put her away. And the woman which hath an husband that believeth not, and if he be pleased to dwell with her, let her not leave him.* (I Corinthians 7:12-13)

Paul is speaking to the believing spouse married to an unbeliever. Why? Just like we saw with Mark 10:2-10 and Matthew 19:3-9, Jesus could not reference the law to say when it is right or wrong to put away an unbelieving spouse.

Paul's answer is that it was up to the believing spouse whether they wanted to put the unbelieving spouse away. Paul's actual answer is to the church and he is saying the church shouldn't force a believer to put away their unbelieving spouse.

Everything we have covered relative to divorce and putting away still applies to our present time.

CHAPTER 11

Church

GOD CREATED BOTH marriage and church as a way for us to be profitable and generate the spiritual value needed to bring about God's will! We have seen the importance of following God's word: it is profitable and leads us to facilitating God's will. The previous five chapters covered marriage, so now it's time to cover church.

We saw the Bible ended with believers occupying the new Jerusalem, which was called the tabernacle of God. This can be seen as the church. We also saw the new Jerusalem was referred to as the Bride of Christ. This can be seen as marriage.

The meaning of life is church and marriage. Notice, both of these are group wills. Both of these are supposed to generate spiritual value for everyone involved, through the interaction of the individuals in their uniqueness, with the right and just Holy Spirit flowing through them by grace.

The first area to cover is the purpose/mission of church. It turns out, there is a lot of confusion about this. For instance, is church a business, a ministry, or both? In order to understand the role of the church, let's contrast a business with a ministry.

Business or Ministry

The goal of a business is to make money through offering a good or a service. The measure of success is how many resources have been accumulated. Example: The goal of a hardware store is to make money through selling household hardware for home improvement. A successful hardware store has a lot of employees and a lot of money.

The goal of a ministry is to offer goods or services through resources (including money). The measure of success is how well people have benefited from the goods or services that were offered. Example: The goal of a soup kitchen is to feed as many people as possible with a healthy meal through accumulating resources (food). A successful soup kitchen feeds a healthy meal to everyone who needs a meal that lives within the vicinity of their ministry.

Which one is church? More to the point, which one results in generating spiritual value for everyone involved through the interaction of the individuals in their uniqueness with the right and just Holy Spirit flowing through them by grace?

If a church measures its success by the amount of money it has raised and the number of people in the church, they are a business. Measuring success in church this way is commonly referred to as a focus on "nickels and noses." If a pastor feels good about his church because he has more members or a bigger budget than another church, he is testifying that he sees church as a business. If his response to "How's your church doing?" is to talk about attendance and tithes, then he sees church as a business.

There are books that not only see the pastor as a CEO, they see Jesus as a CEO! Worse, they see the Bible as their handbook on how to make money. These churches are focused on getting money from their members, and they do it by explaining why you need to tithe so the pastors get paid and the building, which is what they consider the church and not the people, needs to be beautiful. The deception is they equate the amount of resources they have accumulated with the quality of their message and whether they are in God's will. It's as if they're thinking, "We must be in God's will because we have a lot of people and money."

What was Jesus' response to having thousands of followers?

> *Then Jesus said unto them, Verily, verily, I say unto you, Except ye eat the flesh of the Son of man, and drink his blood, ye have no life in you.* (John 6:53)

This drove away thousands of followers. Jesus was not focused on quantity, He was focused on quality.

If a church measures its success by the quality of help they have offered people, they are a ministry. Notice, this includes people inside and outside of the church, which is the community of believers. If a pastor feels good about his church because he can reference people who have been helped in a lasting way, he is testifying that he sees church as a ministry. If his response to "How's your church doing?" is to talk about the health of the community, then he sees church as a ministry.

Notice, a ministry requires resources, including money, in order to achieve its goal. If someone denies the fact that the church needs money and resources (laborers) in order to provide a benefit, then that person is not truly a part of the community. How would you feed the homeless without resources?

A ministry sees money and resources as a hurdle. This means there is a minimum amount needed to deliver the benefit, and the leader would be able to identify and objectively measure the benefit. If this isn't addressed, the community will go off course or get distracted.

A ministry sees the services as the driver: helping people, both inside and outside the community, get out of the hole and grow in God is leadership and love. This can only be done with God's doctrine.

It takes money to do this. Next to the kingdom of God, Jesus spoke about money more than any other topic. Why? Because the principles it takes to accumulate spiritual value are the same principles it takes to accumulate physical value! You must be wise with your resources, especially when it comes to investing, whether it is mutual funds or prayer. Ultimately, you ought to be investing as much of your spiritual value in drivers instead of hurdles.

Christians have a responsibility to partner with God in providing a service to people, while being responsible with money to the same extent they would be responsible with spiritual value. However, the reason the church has become a business is that Christian leaders can't explain or demonstrate the benefits of Christianity. So far, we have seen the benefits are salvation and sanctification.

> *But seek ye first the kingdom of God, and his righteousness; and all these things shall be added unto you.* (Matthew 6:33)

Jesus said to focus on the intangible, while the tangible resources would be provided by God as an effect. Leaders who focus on tangible resources as a cause are hindering God's will. Ultimately, a business is focused on getting, while a ministry is focused on giving, which is love and leadership instead of bossing.

Ask a pastor what benefit they are offering to people. Ask them if they would provide that benefit even if they weren't getting paid a salary. Today, we are hindering God's will through one-half of the way God provided to bring about His will because people can't identify the specific benefit, and they flip causality with respect to Jesus' words. Let's look at the standard church in order to understand causality more clearly.

Acts Church

The standard that every church in our dispensation is contrasted with is "the Acts church," which was the church that formed immediately after Pentecost and was described in Acts 2. In Book 1, we stated it was the group example of being blessed by doing the opposite of the four causes of judgment. Here is a quick summary of what we've already covered.

> *And fear came upon every soul: and many wonders and signs were done by the apostles. And all that believed were together, and had all things common; And sold their possessions and goods, and parted them to all men, as every man had need. And they, continuing daily with one accord in the temple, and breaking bread from house to house, did eat their meat with gladness and singleness of heart, Praising God, and having favour with all the people. And the Lord added to the church daily such as should be saved. (Acts 2:43-47)*

This is the famous passage about the ultimate New Testament church. This church is a rarity today. The most asked question I get from people is, "Why don't we see signs and wonders in churches today? Wouldn't that help church growth?" Basically, they are asking why can't we experience the Acts church today?

Acts 2:43-47 lists the effects (*what*) that some people wish we had in church today: signs, wonders, and the Lord adding to the church daily people who are receiving salvation.

We saw in *Modeling God* that love is "the giving of a value to someone and not expecting anything in return from that person." The effects that were listed were love towards each other. In fact, it looks like the desired effects were Jesus' second command ("Love your neighbor")! What were the causes of the Acts church? We saw the preceding verse (Acts 2:42) gave the causes:

> *"And they continued stedfastly in the apostles' doctrine and fellowship, and in breaking of bread, and in prayers."*

There were four causes to getting the effects of the Acts church:

- Apostles' doctrine
- Fellowship
- Breaking of bread
- Prayers

Do these causes look familiar? How many of these causes do you see in today's churches? Notice, in Mark 12:30, Jesus said the following when explaining the first command:

> *"And thou shalt love the Lord thy God with all thy heart, and with all thy soul, and with all thy mind, and with all thy strength: this is the first commandment."*

In this passage, mind was G1271 dianoia. This word is made up from two words G1223 dia and G3563 nous. The latter root word is the word mind (or soul). The former root word means the channel through which it flows. So, we have Jesus explaining a physical organ through which the mind flows, which we know to be the brain. Jesus gave a word for brain, and an explanation for how the mind and brain interact before the translators of the King James Version even understood this!

Jesus listed four parts. Actually, every person is made up of these four parts:

- Heart – Emotion
- Soul – Spirit
- Brain – Mental
- Strength – Physical

We saw that Paul referred to the New Testament church as the body. Now look at the four causes of the Acts church! Acts 2:42 addressed all four areas of the New Testament church (body):

- Mental – Apostles' doctrine
- Emotion – Fellowship
- Physical – Breaking of bread
- Spirit – Prayers

Then one of them, which was a lawyer, asked him a question, tempting him, and saying, Master, which is the great commandment in the law? Jesus said unto him, Thou shalt love the Lord thy God with all thy heart, and with all thy soul, and with all thy mind. This is the first and great commandment. And the second is like unto it, Thou shalt love thy neighbour as thyself. On these two commandments hang all the law and the prophets. (Matthew 22:35-40)

Loving the Lord thy God will result in loving your neighbor. Jesus' first command is a cause and the effect is the second command. Jesus' first and second command are *one* because they are both halves of causality (cause and effect). Is it surprising that the same principle would be true for the New Testament church (body) as it is for the New Testament individual body?

- The first command is faith.
- The second command is grace.
- All of it is made perfect because of love.

Loving the Lord thy God with all your heart, mind, soul, and strength is faith because it requires a belief in something you can't see (God) and something that hasn't happened yet (giving before receiving). The effect is that we love our neighbors: the second command. It is not in our nature to love. The only way to do this is for God to do it through us, which is grace ("the divine influence upon the heart, and its reflection in the life"). The second half of that definition is the focus; it is God doing it through us by the Holy Spirit so that it is reflected off of us. It is not from us, and we can't take credit.

The proof and profitability of the second command is dependent on the second half of the definition of grace. When a person doesn't understand this second half or doesn't think it needs to be mentioned, they are clearly missing the proof and profitability of being a believer, which is love.

> *By this shall all men know that ye are my disciples, if ye have love one to another.* (John 13:35)

Jesus said that our love for one another proves to *all* men that we are His disciples. Again, love is the ultimate effect of God flowing through us. God measures performance in terms of profitability. These are the effects. Whether it is the individual doing the first command or the New Testament church; when the body is operating in all four areas, then the effects of a believing body are seen.

Acts 2:42 is how the church collectively does Jesus' first command! The effects God wants to see in church are Acts 2:43-47; the church collectively doing Jesus' second command! This draws in people to get saved!

Some pastors and teachers preach about the effects because they don't know the causes. Anyone who tells you to behave a specific way (Acts 2:43-47) without explaining the cause (Acts 2:42) is preaching the effects. If leaders want to talk about effects, what is the ultimate effect leaders should be focusing us on?

Church Leadership

And Jesus came to them and spake unto them, saying, All authority hath been given unto me in heaven and on earth. Go ye therefore, and make disciples of all the nations, baptizing them into the name of the Father and of the Son and of the Holy Spirit: teaching them to observe all things whatsoever I commanded you: and lo, I am with you always, even unto the end of the world. (Matthew 28:18-20, ASV)

These were Jesus' last words to the disciples. Since the disciples represented the church, these were Jesus' last words to us. We said salvation and sanctification were the benefits the ministry provided. Now we see the ultimate benefit: disciples. In the verse above that covered love, Jesus stated people will know we are His disciples by our love for one another. The goal is to be a disciple. What are disciples?

Disciples are leaders that not only can help people grow in salvation and sanctification, they are able to grow people to be able to grow others in salvation and sanctification. They are leaders who make leaders. It looks as if we are far enough from the wall to see the image of a person from the right side of the wall teaching the person who is themselves teaching the original person to fish!

This replication is called generativity. This isn't just sustainability. This is the plan for how the entire earth would have been saved! Making disciples is God's perfect will! Are you in God's perfect will?

Do you think it's possible Jesus' first words to us will be: "Did you do what I told you to do to bring about God's perfect plan? Or did you do a man-made plan based on being a boss instead of a leader?" Remember, everyone can be a leader, but only one person can be the boss. Also, leaders like to be led, but bosses don't like to be bossed.

Is your church making disciples? Why or why not?

Conclusion

God will never achieve His perfect plan (all are saved) because people chose not to come unto the knowledge of the truth. We know God's ultimate plan definitely occurs, but we determine when and at what cost.

God's ultimate plan: The meaning of life is church and marriage. Those who are Righteous by His grace are the church living in the new Jerusalem on the new earth married to Jesus who paves the way for that Bride to fellowship with the Father for eternity. This is also how God intended to bring about this plan: church and marriage!

God created both!

Church ought to be a group of people who interact in the uniqueness God created them to be so that everyone ends up with more spiritual value. The ultimate way to do that is to have the Holy Spirit flowing through each person without any hindrance from each person. This is sanctification. If this is done towards other people, that would facilitate replication.

Marriage ought to be two people who interact in the uniqueness God created them to be so that each person ends up with more spiritual value. The ultimate way to do that is to have the Holy Spirit flowing through each spouse at the greatest possible amount toward the other spouse.

The people participating in this type of church and this type of marriage would accumulate spiritual value that allows God to move, facilitating God's will, while also facilitating God's plan by allowing the Holy Spirit to move through them, as an individual and a group, into this world.

Jesus' last command was to make disciples, which can be seen as God's perfect will. That is the ministry of the church!

Dr. Joel Swokowski's Commentary

The church my family attended when I was a teenager was the largest church in the area. We're talking thousands of people in attendance every week. It was a church that started as a street ministry, evangelizing in the local pubs, and grew to what I saw as a teen.

Years later, I became a member as an adult, no longer because my parents attended but because I wanted to grow in the grace and knowledge of the Lord. I joined their discipleship program and was blessed to be paired up with the senior pastor as my small group leader. Within a year, I left that church, largely due to the abuse I witnessed at the hands of the senior pastor and his elders.

Not long after leaving that church, I attended a wedding that was held there. Wouldn't you know, the senior pastor was there! I ran into him in the lobby and asked him, "How's the church going?" Even then, I was struck by the answer he gave. He said things were going great and added that their attendance and tithes were up. He measured the success of his church through tangible and quantitative measures (nickels and noses).

He's not alone. I think this happens a lot, especially with senior pastors who end up having to be too focused on the logistics of having a building rather than on caring for people. It's actually made me grateful my community hasn't grown into a megachurch. I want to care for as many people as possible in a manner that creates disciples, but I want that to happen in a way that facilitates me to grow my focus on the mission and people, not on the resources.

As a teaching pastor, I've always been interested in the approach other churches and pastors take to help their members understand the importance of tithing. It tends to be a touchy subject that pastors prefer not to preach, yet, they know it's important.

The founding senior pastor of the church I pastor is my dad. He played a huge role in the lives of the leaders of our church, bearing the weight of senior pastor due to his having decades of experience. He preached the only sermon I remember hearing preached on tithing in over ten years! There are two reasons for this:

1. Our church has a great understanding of justice and reward. They see tithing as a cause to invest into the church, not like a bill they have to pay.

2. His sermon emphasized the spiritual (intangible cause) benefit and responsibility of tithing. He likened tithing to our ability to care for the Bride of Christ while Christ is away, preparing a place for her. This is an incredible analogy that emphasizes the spiritual investment that tithing can and ought to be.

It's not wrong to teach tithing. Just as with any topic, it's more about the *how* and *why* behind *what* is being taught than merely *what* is being taught. What our pastor taught us is that tithing is about investing into people, just like leadership. It's about people, not a building! It's about people, not a paycheck!

Does your church see tithing as an investment into people?

Does your church see tithing as a way to care for the Bride of Christ?

Laborers

There is a role within the church that has been left by the wayside, and it's pivotal to all of our mental and emotional health. I have said that the church has contracted out its responsibility over our mental and emotional health to psychology. I think this is due to the church losing its ability to provide this benefit. Were psychologists mentioned in the Bible? Why don't we read stories of our Bible heroes going to therapy?

It's because God provided a way for the church to fill this need, and it was done through the role of laborers. The following is from Matthew 9:35-38.

> *And Jesus went about all the cities and villages, teaching in their synagogues, and preaching the gospel of the kingdom, and healing every sickness and every disease among the people.* (Matthew 9:35)

This verse sets the context for the understanding we need. Take a moment and realize everything this verse stated: Jesus healed *all* manner of disease and *all* manner of sickness. Jesus had done every possible miraculous healing He could have done for these people.

> *But when he saw the multitudes, he was moved with compassion on them, because they fainted, and were scattered abroad, as sheep having no shepherd.* (Matthew 9:36)

In my experience, most people miss the impact of verse 36 because they focus on the people being sheep having no shepherd, thinking that all these people needed was a pastor. While that was part of this verse, the part that's most important tends to be skimmed over. Let's look at what *fainted* and *scattered* meant.

The word fainted was Strong's #1590 eklyo — "to loose, unloose, to set free." The etymology of this word points towards: to stop covenant, to stop agreement, to disassemble. We can see these people were loose from each other. These people were essentially individuals without any connection to each other, which prevented agreement. People can be gathered together yet still isolated from each other.

The word scattered ("scattered abroad") was Strong's #4496 rhipto — "to cast down, throw." The etymology of this word points towards: harassed in thought, mental dejection, thrown in different directions affecting the ability to focus and concentrate. The people had a bad thought process.

Summary: The people were *physically* healed; however, they weren't connected to each other (they were individuals), and they had a bad thought process. When we remember that people can instantly get spiritual healing from God, this brings up an uncomfortable conclusion:

While God and Jesus can immediately and miraculously heal people spiritually and physically, God and Jesus are unable to immediately and miraculously heal a person's thought process (mental) or cause people to be in unity (emotional).

Let's continue (we haven't even gotten to laborers yet!!). It's important to remember, Jesus had miraculously healed these people of every physical ailment, yet the people still needed mental and emotional healing.

Now what Jesus meant makes sense when He said they were like sheep without a shepherd. What ought a shepherd (pastor) do?

Two requirements:

1. Help people have a good thought process.
2. Help people come into agreement through fellowship.

How are today's religious leaders doing relative to these two requirements?

Again, these people were physically healed and followed Jesus. For all intents and purposes, we can see these people as believers; yet, they had a bad thought process and were not in agreement. Jesus Himself stated that these sheep (believers) were without a shepherd, a person who *could* help them get mental and emotional healing.

Jesus wasn't done, we still haven't seen the word laborers used!

> *Then saith he unto his disciples, The harvest truly is plenteous, but the labourers are few;* (Matthew 9:37)

Jesus said to His disciples that the harvest is plenteous. A harvest only occurs once there is something to reap. This points to these people being believers. If these people weren't believers, they would be referred to as seeds.

Many people interpret the word laborers as being evangelists, as if the people whom Jesus had compassion on were in need of being witnessed to, as if they needed to be saved. This passage was *not* speaking about evangelists. Is the church really lacking evangelists? It seems to me the main thing the church can agree on is the need to get the message of salvation out to the lost; yet, the church is more divided today than ever before, and the mental/emotional health of the church has been on a steady decline for decades.

Jesus said the harvest is plenteous, but the laborers were few...and our only solution is to tell people they need to receive Christ as their Lord and Savior?!?

Jesus *needed* laborers and this verse came immediately after He stated these people were sheep without a shepherd because they had a bad thought process and were not in agreement. It seems the shepherd (pastor) is responsible for ensuring the people in his care are given laborers.

Now we see the job of the laborers was to provide mental and emotional healing by:

1. Help people have a good thought process.
2. Help people come into agreement through fellowship.

Jesus *needed* laborers for two reasons:

1. This was *not* something He could miraculously accomplish.
2. This was something that was going to take *daily* work.

Repairing thought processes and building agreement are both attributes that take work every day.

Let's look more closely at the word laborer.

The word laborer was Strong's #2040 ergates — a workman, a laborer. The etymology of this word pointed towards the same word that was translated as workman in II Timothy 2:15:

> *Study to shew thyself approved unto God, a workman that needeth not to be ashamed, rightly dividing the word of truth.*

The word laborer/workman meant to toil with works of words as a teacher.

Jesus needs people who *daily* toil with works of words as a teacher in order to improve the thought processes of believers and help the church come into agreement through fellowship.

Let's finish this section with verse 38.

> *Pray ye therefore the Lord of the harvest, that he will send forth labourers into his harvest.* (Matthew 9:38)

Jesus told His disciples to pray that God would send forth these laborers/workmen into His harvest. Again, not only did this speak about believers, but Jesus *Himself* stated prayer was needed to fulfill this need. If this was something that Jesus could miraculously do, He would *not* need us to do it, let alone tell us to pray for these laborers to be sent forth.

Summary

- God and Jesus are unable to miraculously heal a person's thought process or cause people to be in unity.
- God is responsible for restoring a person spiritually and physically.
- We are responsible for restoring ourselves and others mentally and emotionally.
- Laborers help a person be restored mentally and emotionally.
- Laborers ought to take direction from God when doing the daily and hard work in helping others, but it is still the laborers (our) responsibility to provide this benefit to the church.
- Ultimately, disciples (and the church) ought to restore people mentally and emotionally.

CHAPTER 12

Church Leadership

JESUS' LAST COMMAND proved the church is a ministry that provides disciples in order to accomplish God's perfect plan relative to individuals. When it comes to God's group plan, Jesus provided gifts to grow the health of the church in order to support His last command. Here's the explanation from Paul:

> *And he gave some, apostles; and some, prophets; and some, evangelists; and some, pastors and teachers; For the perfecting of the saints, for the work of the ministry, for the edifying of the body of Christ: Till we all come in the unity of the faith, and of the knowledge of the Son of God, unto a perfect man, unto the measure of the stature of the fulness of Christ: That we henceforth be no more children, tossed to and fro, and carried about with every wind of doctrine, by the sleight of men, and cunning craftiness, whereby they lie in wait to deceive; But speaking the truth in love, may grow up into him in all things, which is the head, even Christ: From whom the whole body fitly joined together and compacted by that which every joint supplieth, according to the effectual working in the measure of every part, maketh increase of the body unto the edifying of itself in love.*
> (Ephesians 4:11-16)

Let's look at the five gifts Jesus provided.

Apostle

Apostle: The word in Ephesians 4:11 was G652 apostolos.

> G652 apostolos — "a delegate, messenger, one sent forth with orders" from #649
> G649 apostello — "to order (one) to go to a place appointed" from #575 and #4724
> G575 apo — "of separation" (We saw this with "putting away": apolyo)
> G4724 stello — "to set, place, set in order, arrange"

An apostle is a person who leaves one place and gets sent to another place in order to arrange and set in order a new community. In the diagram below, the vertical lines to the left represent the people in a church. One of those vertical lines (apostle) was supposed to go to a new place (arrow) to begin a new church. The dots are unbelievers in this new area.

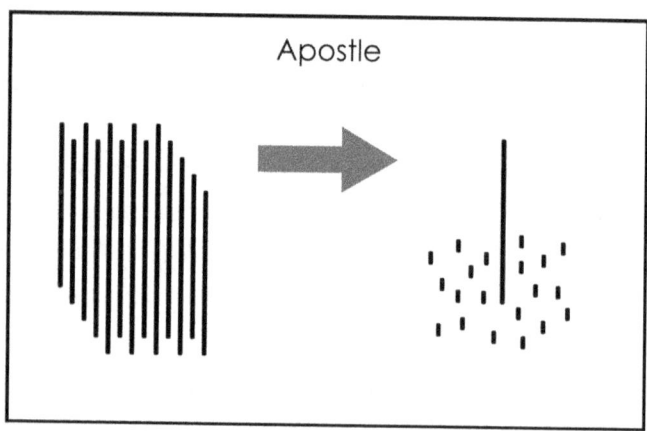

> *For I think that God hath set forth us the apostles last, as it were appointed to death: for we are made a spectacle unto the world, and to angels, and to men. We are fools for Christ's sake, but ye are wise in Christ; we are weak, but ye are strong; ye are honourable, but we are despised. Even unto this present hour we both hunger, and thirst, and are naked, and are buffeted, and have no certain dwellingplace; And labour, working with our own hands: being reviled, we bless; being persecuted, we suffer it: Being defamed, we intreat: we are made as the filth of the world, and are the offscouring of all things unto this day. I write not these things to shame you, but as my beloved sons I warn you. For though ye have ten thousand instructers in Christ, yet have ye not many fathers: for in Christ Jesus I have begotten you through the gospel.* (I Corinthians 4:9-15)

Apostles are fathers in the faith because they begin the work of the church. They pave the way for the saints by taking the abuse. Paul listed the abuse he received as an apostle in II Corinthians 11, which included being given thirty-nine lashes five times, beaten with rods three times, shipwrecked three times, and stoned once. We can even see people in the Old Testament acting as apostles: Abram, Joseph, Moses, Daniel, etc.

Apostles are truly leaders because not only do they not receive an immediate benefit, they receive immediate abuse. This was true for the original twelve Apostles and for all the apostles that came after. Remember, the church is made up of saints. Strong's Concordance gives the definition of saint as G40 agios - "most holy thing." The implication is that holy things are set apart, so saints are also seen as the called out ones. By contrast, the apostles are the sent out ones.

Prophet

Prophet: The word in Ephesians 4:11 was G4396 prophetes.

G4396 prophetes — "one who moved by the Spirit of God and hence His spokesman, solemnly declares to men what he has received by inspiration, especially concerning future events, and in particular such as relate to the cause and kingdom of God and to human salvation" from #4253 and #5346

> G4253 pro — "before"
> G5346 phemi — "say"

Clearly, a prophet is supposed to say something before it actually happens. In the Old Testament, only the prophets had the indwelling Holy Spirit. Consequently, they were able to explain God's salvation as well as coming events. Notice, the Strong's definition could be applied to any believer today because of the indwelling of the Holy Spirit. This is why I've represented the prophet with a tube in the shape of an "L." God's grace would flow into the top of the tube and out to all the people.

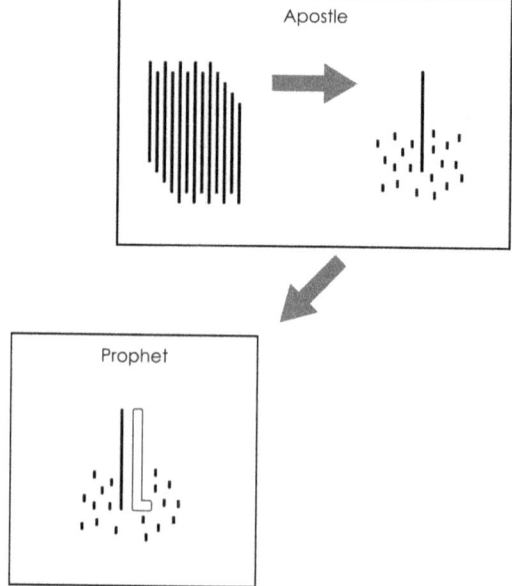

Prophets need to be master communicators. They need to make sure they aren't adding to God's message and they need to make sure they aren't hindering God's message. The confusion with prophets today occurs because people don't understand leadership.

In the Bible, there was only one instance where a prophet stated he was a prophet:

> *So Ahab sent unto all the children of Israel, and gathered the prophets together unto mount Carmel. And Elijah came unto all the people, and said, How long halt ye between two opinions? if the Lord be God, follow him: but if Baal, then follow him. And the people answered him not a word. Then said Elijah unto the people, I, even I only, remain a prophet of the Lord; but Baal's prophets are four hundred and fifty men.* (I Kings 18:20-22)

Notice, this statement of being the only prophet of the Lord didn't immediately benefit Elijah. If anything, it was a death sentence. The bottom line is Elijah did not facilitate his purpose and progress with this statement. People today who claim to be "prophets" while facilitating their own benefit are proving they are not prophets of God.

In the Bible, a prophet never benefited from their prophecy. Most of the prophecies recorded in the Bible were fulfilled after the prophet died. Even when prophets could benefit, they turned down payment. An obvious example of this concerned Elisha. He refused to benefit from healing Naaman from leprosy (II Kings 5:6).

Apostles and Prophets

For now, notice the Bible showed the importance of the apostles and prophets, especially how they worked together to build the foundation for the new Jerusalem through communication (Ephesians 2:19-22).

Likewise, the next verse made the same point for the current churches:

> *Now ye are the body of Christ, and members in particular. And God hath set some in the church, first apostles, secondarily prophets, thirdly teachers, after that miracles, then gifts of healings, helps, governments, diversities of tongues.* (I Corinthians 12:27-28)

This verse from Paul was written after Pentecost to a church in a letter that we use today to guide how the current church ought to operate. A biblical church ought to be built on the work of an apostle (who began the work) and a prophet (who has a message from God). The apostle ought to be able to take abuse and the prophet ought to be able to focus on benefiting others.

Some people think that there weren't any apostles other than the first twelve. What about Paul? Some people then state that Paul was the only apostle other than the twelve. What about this verse?

> *Which when the apostles, Barnabas and Paul, heard of, they rent their clothes, and ran in among the people, crying out,* (Acts 14:14)

Some people then state they think Jesus took these two gifts away after Baranabas and that we don't have apostles and prophets today. Look at this verse from Jesus' mouth to a church that existed after Paul, Barnabas, and all of the twelve original Apostles except the one who wrote this book were killed:

I know thy works, and thy labour, and thy patience, and how thou canst not bear them which are evil: and thou hast tried them which say they are apostles, and are not, and hast found them liars: (Revelation 2:2)

If apostles could only be the original twelve Apostles (and also included Paul and Barnabas) and none of them were available to this church, why would a church have to test apostles? The word tested was G3985 peirazo — "to try whether a thing can be done." Is it really a test to find out if the person was one of the Apostles?

Remember, the apostles and prophets don't receive a benefit; they receive persecution. If you don't believe there are apostles and prophets today, take a moment to state your reason out loud or write it down before continuing.

The church today ought to be built on the foundation of the communication from God delivered by the apostle and the prophet. The apostle and prophet would exhibit mada under testing conditions to the point others would recognize it. The next step in Jesus' strategy for growing the church focused on communicating this message from God to others.

Evangelist

Evangelist: The word in Ephesians 4:11 was G2099 euaggelistes

> G2099 euaggelistes — "a bringer of good tidings" from #2097
> G2097 euaggelizo — "to bring good news" from #2095 and #32
> G2095 eu — "to be well off"
> G32 aggelos — "a messenger"

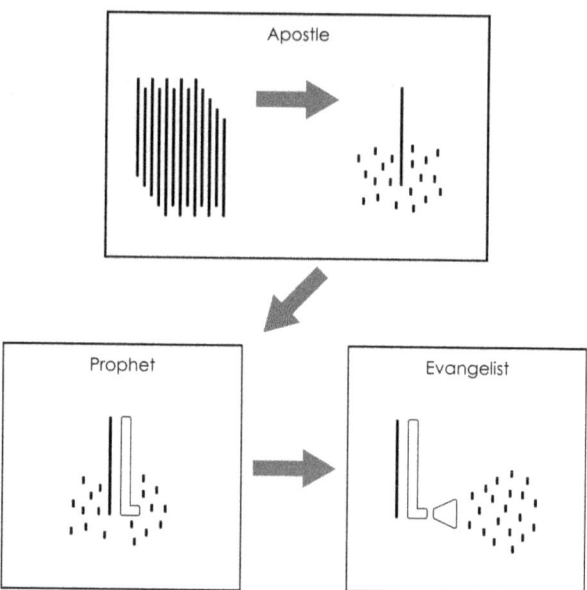

An evangelist brings a good message out to others. A "herald" who amplifies the message also does this. In our diagram, the evangelist is an amplifier (trapezoid on its side) drawing all the dots towards the apostle and prophet.

Pastor

Pastor: The word in Ephesians 4:11 was G4166 poimen

G4166 poimen — "a herdsman, especially a shepherd"

The shepherds work together to feed and protect the people who have been drawn in. This was the same word used by Jesus when He spoke about Himself:

I am the good shepherd, and know my sheep, and am known of mine. As the Father knoweth me, even so I know the Father: and I lay down my life for the sheep. (John 10:14-15)

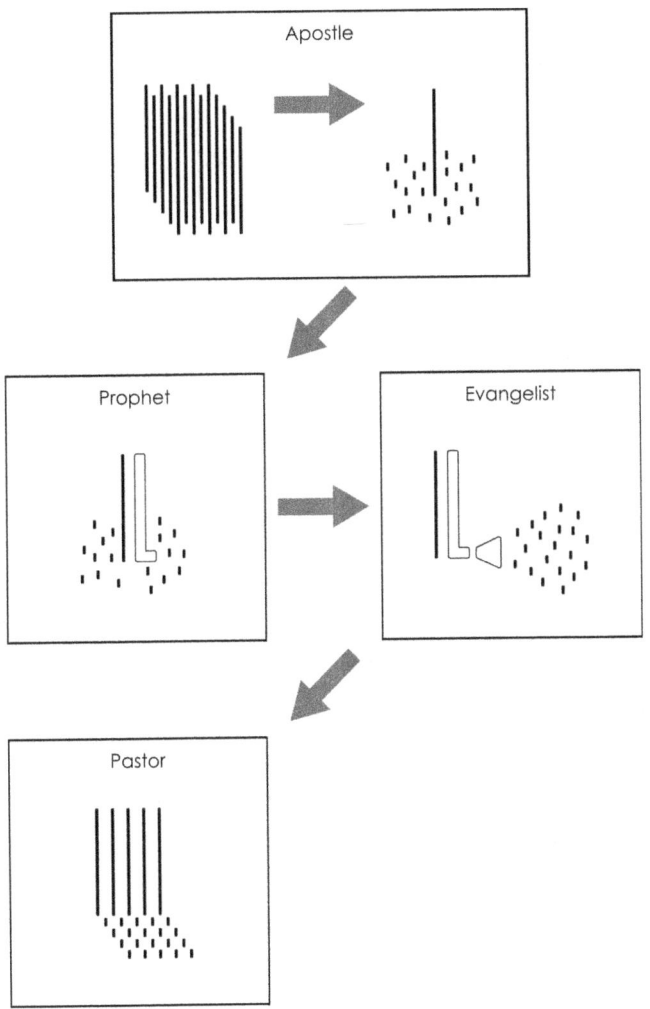

Clearly, the pastor is a leader. However, shepherds actually work in groups when they are handling sheep. Likewise, it looked as if the Bible recognized many shepherds being led by the head shepherd. In the Old Testament, we saw a group of priests ministered to the people and one priest was identified as the high priest. We represented this step as shepherds (lines) organizing the flock (dots in rows).

In the New Testament, it looked like the group of shepherds was translated as *elders*. Here are two passages for you to decide which of the five gifts Jesus provided to the church would have the responsibilities of *elders*:

> *Then the disciples, every man according to his ability, determined to send relief unto the brethren which dwelt in Judaea: Which also they did, and sent it to the elders by the hands of Barnabas and Saul.* (Acts 11:29-30)

> *And there came thither certain Jews from Antioch and Iconium, who persuaded the people, and having stoned Paul, drew him out of the city, supposing he had been dead. Howbeit, as the disciples stood round about him, he rose up, and came into the city: and the next day he departed with Barnabas to Derbe. And when they had preached the gospel to that city, and had taught many, they returned again to Lystra, and to Iconium, and Antioch, Confirming the souls of the disciples, and exhorting them to continue in the faith, and that we must through much tribulation enter into the kingdom of God. And when they had ordained them elders in every church, and had prayed with fasting, they commended them to the Lord, on whom they believed.* (Acts 14:19-23)

This last passage showed Paul was a true apostle: He got stoned for preaching and when he didn't die, he returned to keep preaching!

The passage also showed once the apostles (Paul and Barnabas) had a group of believers, they appointed elders in every church. This raises an interesting point: apostles appoint elders/pastors. Why is that?

The elder/pastor is the visible leader of the church. If the elder/pastor campaigns or applies for this job, aren't they facilitating their own purpose and progress? Doesn't that prove they are not a leader? Also, if an apostle begins the church and then places himself in the role of pastor, doesn't that prove he is not a leader? However, the apostle is like John the Baptist, while the pastor is like Jesus. The apostle points people to the shepherd.

How many issues do we have today because the pastor of a church is the boss? The pastor is the final authority, and he makes sure he immediately benefits. What if the authority over a church was an apostle who was not present and the leader was the pastor who was present? The apostle couldn't benefit at all from being in charge of the church, and he would intentionally point the people to the pastor. Meanwhile, the pastor would be a leader who had accountability. The members could always reach out to the apostle if they have an issue with the pastor. This is a dissolve approach and it is God's plan for church!

This was exactly how Paul handled the churches he fathered! Paul was the apostle over the churches, but he wasn't present. Instead, he appointed elders/pastors over the church. Paul wrote this to the church in Corinth:

> *If I be not an apostle unto others, yet doubtless I am to you: for the seal of mine apostleship are ye in the Lord.* (I Corinthians 9:2)

First, notice that Paul could be an apostle to one group and not another. This speaks to the responsibilities of beginning the church and not some universal title. Second, we've seen God's plan for governing involves a leader and a wise person. In the Bible, apostles and prophets served as the wise person, not the leader. Melchizedek was the priest during Abram's time. Aaron was the priest during Moses' time. Pharaoh and Nebuchadnezzar/Darius were the kings during Joseph's and Daniel's times, respectively.

> *Therefore also said the wisdom of God, I will send them prophets and apostles, and some of them they shall slay and persecute:* (Luke 11:49)

This is just one verse that reinforced that the apostles and prophets don't receive a benefit; they receive persecution! Why? Because their job is to confront leadership!

The apostles and prophets could do this because they never directly benefited from their job.

However, almost all of today's churches do not have the benefit of these two gifts from Jesus. People driven by an animal thought process have removed the limitation from the conjunctive definitions of apostle and prophet. The result? Apostles and prophets directly benefit from their office, usually with money.

People driven by a human thought process have increased the limitations in the conjunctive definitions of apostle and prophet to the point they believe Jesus took these two gifts back, and there is currently no one who can confront and correct the pastors in their leadership role. Do you believe it's God's will to have apostles and prophets in the church today? Do you believe people against apostles and prophets are in God's will?

Worse, Ephesians 4:13 stated Jesus gave these gifts *till we all come in the unity of the faith, and of the knowledge of the Son of God, unto a perfect man, unto the measure of the stature of the fulness of Christ.* Notice, we have not attained this. Did Jesus fail?

In fact, the word for gave was G1325 didomi and interlinear Greek translations use the word gives because the verb was third person singular, indicative Aorist active. Basically, the subject was Jesus from a third person singular perspective. The indicative Aorist means it was an event that happened in the past, with the active voice indicating the action continues today.

Jesus did the cause in the past that provided and continues to provide and will continue to provide until the objective is met. The same presentation was given when talking about Jesus providing salvation: He did something in the past (was crucified) that provided for salvation and continues to provide for salvation today.

Notice, people who think there are no apostles and prophets today either believe: we have all attained the unity of the faith and of the knowledge of the Son of God, Jesus didn't keep His word, Jesus' power ran out, or Jesus' ability to continually provide for salvation can also be taken back or run out. Which one do you believe?

It looked as if the first three gifts from Jesus were people who were called by God and didn't immediately benefit from their position. This fourth gift (pastor/elder/shepherd), given to feed and protect the people, was the first gift appointed by another person and could benefit from the people. For example, they could receive a salary.

Teacher

Teacher: The word in Ephesians 4:11 was G1320 didaskalos

G1320 didaskalos — "a teacher" from #1321
G1321 didasko — "to teach"

Teachers are supposed to help everyone grow in maturity, which is represented as lines in the diagram. It is from here that this church can send out an apostle, and the process can continue indefinitely in a generative manner.

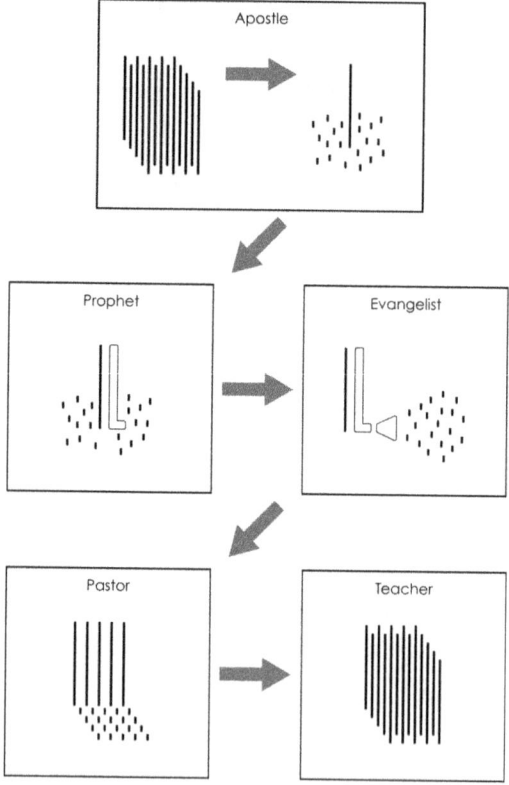

Summary

What will we say if Jesus asks us why we didn't follow His plan for growing the church on earth? Are we doing a man-made plan that hinders God's plan?

Today, how many churches were begun by an apostle and then pastored by a different person? How many churches have an apostle who doesn't benefit from having begun the church? Is it possible to replicate this biblical plan for beginning churches if we don't believe apostles and prophets exist today? How ought pastors to be confronted? Do you believe pastors are never wrong?

Now we can see the previous representation of God flowing through a group of people on the right half of the wall is church. God is flowing through these people so they exchange with each other in their uniqueness. We can see this group of people responsible for the person teaching the person who is teaching the original person how to fish.

God created both marriage and church as a way for us to be profitable and generate the value needed to bring about God's will! We have seen the importance of following God's word: it is profitable and leads us to bring about God's will.

The last seven chapters have shown how man's tradition has been put in place of God's doctrine of marriage and church. How will you respond to people who are hindering God's will by defining marriage and church according to man's tradition?

Let's begin bringing everything to a conclusion by looking at an area where man's doctrine greatly differs from God's doctrine as presented in His word.

Dr. Joel Swokowski's Commentary

Five-Fold Memory Device

A simple way to remember the spiritual gifts presented in Ephesians chapter 4 is by using your hand. Go ahead and hold your hand up for me and go through the following steps:

1. Hold out your thumb as if you're hitchhiking. Now say, "Apostles are always on the road."
2. Point your index finger. Now say, "Prophets point things out."
3. Look at your middle finger. Now say, "Evangelists have the longest reach."
4. Hold up your ring finger. Now say, "Pastors are married to the church."
5. Place your pinky near your ear as if you had an itch. Now say, "Teachers are always in your ear."

This memory device was the first method used to teach me these gifts and has stood the test of time. Apostles are on the road starting new things (churches, communities, projects, etc.). Prophets make people aware of effects that will transpire from specific causes. They point out those effects. Evangelists amplify a message, reaching the masses. Pastors are husbandmen to the church. Teachers give information that helps train up the church.

Three Classes of Spiritual Gifts

Modeling God taught about your spiritual ARE. In the "Determining Your ARE" chapter, Lenhart wrote: *"there is a quiz according to*

Romans 12 that will help you get a big-picture direction for determining your uniqueness. For further information, go to: www.modelinggod.com."

The reference to Romans 12 points to another list of spiritual gifts within the scriptures. In addition to Ephesians 4 and Romans 12, there's also I Corinthians 12 to account for. Here's another memory device for you: the three classes of spiritual gifts were each given by a separate member of the Trinity!

Positional Spiritual Gifts

And he gave some, apostles; and some, prophets; and some, evangelists; and some, pastors and teachers; (Ephesians 4:11)

The positional spiritual gifts are the specific benefits that Christ gave and continually gives to the church in order to make a Bride that is equal to Him in stature.

These gifts are meant to be the leadership of the church, daily facilitating the purpose and progress of people. These gifts are often referred to as the five-fold ministry.

Jesus is the Son of God, and although He has brought many benefits (and continues to do so) to the church, He is particularly focused on two things, one big picture, one small picture.

Jesus' one thing at the big picture level: to connect you to God the Father on a daily basis. Jesus does this by being THE Apostle, Prophet, Evangelist, Pastor (Good Shepherd), and Teacher (Rabbi)! No wonder Jesus was able to give these gifts. He was THE example of each!

Jesus' one thing at the small picture level: the five-fold ministry. Jesus is working through these five-fold spiritual leadership positions, a gift to us, to build up the church daily.

Both of these benefits help us today, daily, and forever.

Manifestational Spiritual Gifts

> *Now there are diversities of gifts, but the same Spirit. And there are differences of administrations, but the same Lord. And there are diversities of operations, but it is the same God which worketh all in all. But the manifestation of the Spirit is given to every man to profit withal.* (I Corinthians 12:4-7)

These gifts are from the Holy Spirit. These verses even give the name we use to title this class of gifts: "But the manifestation of the Spirit" - The manifestational spiritual gifts.

Manifestational spiritual gifts are gifts that happen by the Holy Spirit flowing through a believer, for the benefit of the church. These gifts can happen through any willing believer when the Holy Spirit wills it, for the benefit of the body. These are the spiritual gifts given to the church by the Holy Spirit. No one can claim to possess these gifts on a personal level.

Motivational Spiritual Gifts

> *Having then gifts differing according to the grace that is given to us, whether prophecy, let us prophesy according to the proportion of faith; Or ministry, let us wait on our ministering: or he that*

teacheth, on teaching; Or he that exhorteth, on exhortation: he that giveth, let him do it with simplicity; he that ruleth, with diligence; he that sheweth mercy, with cheerfulness. (Romans 12:6-8)

These are gifts given to the church from God the Father, in the womb, to all people, believer or not. Notice, each of them has a qualifier. For example, "diligence" is the qualifier given to the gift of "ruling." These qualifiers prove these gifts can be done wrong and imply they can be misused. These are called motivational spiritual gifts because they are the cause of your motivation, your energy, and your drive. The motivational spiritual gifts are the gifts given to each person that defines their uniqueness.

For as we have many members in one body, and all members have not the same office: So we, being many, are one body in Christ, and every one members one of another. (Romans 12:4-5)

These verses show us these gifts are the key to us being unified as believers. Humanity needs each of these and they work together towards a purpose: sanctification and repair, the two benefits the church is meant to be experts at. Note: some adjustments have been made in the following list of the gifts in order to bring clarity to the meaning behind what is listed in the King James Version in the verses above.

Here is the list:

1. **Perceiver (prophecy)**: wants you to be aware
2. **Teacher**: wants you to understand
3. **Compassion (mercy)**: wants your pain beared
4. **Giver**: wants you to receive a tangible gift
5. **Server (ministry)**: wants to fill a need

6. **Administrator (ruleth):** wants to coordinate a group
7. **Exhorter:** wants you to be excited about the future

We've also taken the list from Romans 12:6-8 and ordered from an intrinsic focus they have, from past to future. The first gift tends to look to the past the most, while the last gift seems to only live in the future. While each gift provides a benefit in and of itself, when all of these are used together, they facilitate fundamental Christianity:

1. Do what God is telling you to do.
2. Confess and repent when you don't.

Another aspect of these gifts is the first two and the final two are predominately accomplished through speaking (speech). The middle three are gifts that involve activity (do). For example, you *show* compassion. Ultimately, these gifts, when taken in the above order, show us the process for a full confession and full repentance.

Notice, the first three gifts represent the confession:

1. **Perceiver:** I know *what* I did (specifically identified) was wrong.
2. **Teacher:** I know *why* (specific causes) I did it.
3. **Compassion:** I don't want to do it again.

The three gifts that focus on the past match:

1. Perceiver is a past, speech gift focused on making people aware of the *what*.
2. Teacher is a past, speech gift focused on helping people understand the *why*.

3. Compassion is a past, do gift focused on bearing emotional pain…not desiring to do it again.

Notice, the next two gifts represent repentance. True repentance involves an act of repair by the confessor. Repentance is a do. The remaining two do gifts match:

4. Giver is a present, do gift focused on giving something to make up for the loss.
5. Server is a future, do gift focused on fulfilling a need to make up for the loss.

Two future speech gifts remain. How do they result in the perfect repentance?

True repair would result in something more for the person I originally sinned against.

6. Administrator is a future, speech gift that coordinates a group towards a goal.
7. Exhorter is a future, speech gift that encourages people.

The reality is, even though the direct offense occurred between two people, the offender is likely to have told others about it in order to justify themselves. For example, "Bill made me so mad, I had to respond. Bill is acting like a baby."

Perfect repentance would involve me identifying all the people I spoke negatively to about the person I sinned against (Bill). Then, I would correct any incorrect information and speak well of the person I sinned against.

Think about it. These other people may not have had an opinion about Bill. After I sinned against Bill, these other people had a negative perspective of Bill.

However, after my perfect repentance, all of these other people would have a positive perspective about the person I sinned against. Bill actually ends up in a better place than before I sinned against him, which is perfect repair.

God desires unity! When these gifts are used together, it brings the benefit of unity to the church!

Unity through uniqueness!

Apostles

Allow me to address the verse most often used to support the belief we no longer have apostles in our age of the church. In the attempt to fill Judas Iscariot's position as an apostle of the early church, Peter put forth the following qualities they would look for:

> *Wherefore of these men which have companied with us all the time that the Lord Jesus went in and out among us, Beginning from the baptism of John, unto that same day that he was taken up from us, must one be ordained to be a witness with us of his resurrection.* (Acts 1:21-22)

This stated that their choice for whoever replaced Judas must be one who had been with them since the baptism of John, who stayed with them during the days of Jesus' earthly ministry, and who saw the resurrected Jesus. These verses are often interpreted to mean that the requirement for all apostles was to have seen Jesus after His resurrection, which would mean that there are no more apostles today. This verse did not say that.

The apostles of that time wanted that to be a requirement for the sake of their ministry to the early believers. Those who could fulfill this early requirement would be the catalyst for replicating disciples going forward. This does not mean that there have not been apostles since that time. Lenhart presented plenty of evidence above that proves apostles are still being given by Jesus to the church today, unless you believe:

- We have all attained the unity of the faith and of the knowledge of the Son of God
- Jesus didn't keep His word
- Jesus' power ran out
- Jesus' ability to continually provide for salvation can also be taken back or run out

Prophets

> *And Jesus answered and said unto them, Take heed that no man deceive you. For many shall come in my name, saying, I am Christ; and shall deceive many.* (Matthew 24:4-5)

Remember what Jesus said about testifying of Himself? If Jesus had stated, "I am the Christ," then His testimony wouldn't have been truth because His *how* and *why* would have been wrong: He would have been facilitating His own purpose and progress. Likewise, the same could be said about anyone stating they are a prophet.

There are several passages that stated the Church began with apostles and prophets. Here is the one Lenhart referenced above:

Now therefore ye are no more strangers and foreigners, but fellowcitizens with the saints, and of the household of God; And are built upon the foundation of the apostles and prophets, Jesus Christ himself being the chief corner stone; In whom all the building fitly framed together groweth unto an holy temple in the Lord: In whom ye also are builded together for an habitation of God through the Spirit. (Ephesians 2:19-22)

CHAPTER 13

Women

IF THERE IS one area where man's doctrine differs the greatest from God's doctrine, I believe it is the topic of women. This chapter presents several passages from God's word on this topic. Please take your time and prayerfully consider what is being presented with each aspect because I believe this represents the area where we are most hindering God's will.

Help Meet

> *And the Lord God said, It is not good that the man should be alone; I will make him an help meet for him. (Genesis 2:18)*

Some people wrongly quote this as "help mate." What does "help meet" mean?

> meet is G5828 ezer - "aid"

Actually, the Strong's reference for help in this passage is the same as Strong's #5828, so it looks like meet adds nothing to this verse.

The word meet in sixteenth century English was defined as fitting, proper, or even perfect. This would make woman the perfect help for a man. While this would seem to apply to the context of this passage (because the animals were not being meet for Adam), I don't have enough information to support that interpretation, although that is not to say it is wrong. Let's look more closely at ezer.

There are two kinds of help a man can get. He can get help from someone lower than him, like an underling. The second kind of help is from someone higher than him, like from God. For instance, Psalm 60:11 stated:

Give us help from trouble: for vain is the help of man.

In this verse, the first help was Strong's #5833 ezrah and was from the same root as ezer. This was help from a superior, especially since the beginning of the verse was asking for help from God. The second help in the verse was Strong's #8668 tesuwquh. This was help from an inferior. Likewise, the end of this verse said this inferior help was from man, not God.

Everywhere that Strong's #5828 ezer was used in the Bible meant help from a superior. So, Genesis 2:18 stated that woman was God's solution to giving the male help from a superior! People who teach that the woman is an inferior help to a male are teaching a man-made doctrine, not the word of God.

Cleave

Therefore shall a man leave his father and his mother, and shall cleave unto his wife: and they shall be one flesh. (Genesis 2:24)

The man shall cleave to his wife. We saw Jesus quote this verse!

> cleave was Strong's #1692 dabaq - "to impinge, i.e. cling or adhere."

Everything about this word stated the woman was the standard (the stable being) and the man was an imposition on her. Even the context implied a man was the property of, or imposition on, his parents until he married, and then he was an imposition on his wife! Some people try to say that a man and woman equally cleave to each other. Keep that in mind as we look at another passage that used dabaq.

> *Thou shalt fear the Lord thy God; him shalt thou serve, and to him shalt thou cleave, and swear by his name. (Deuteronomy 10:20)*

When it comes to humans cleaving to God, which one do you believe is the "superior": humans or God…or do you think we are equal to God in our cleaving? Take a moment and state your will by answering this question.

If a person believes the man cleaving to the woman makes the man superior or equal to the woman, then that same person believes humans are superior or equal to God!

Last Creation

> *And the Lord God formed man of the dust of the ground, and breathed into his nostrils the breath of life; and man became a living soul. (Genesis 2:7)*

And the Lord God caused a deep sleep to fall upon Adam, and he slept: and he took one of his ribs, and closed up the flesh instead thereof; And the rib, which the Lord God had taken from man, made he a woman, and brought her unto the man. (Genesis 2:21-22)

God was limited in what He could make from the dust relative to what He could make from a bone. I realize this may rub people the wrong way who think God could never be limited by anything, but we have dealt with that fact since *Modeling God*. Besides, if God was never limited, then how can things that God makes be better or worse than other things?

I believe God made the perfect being from dust: man. You could not make anything more profitable than "man" out of dust. God is perfect and thinking He could have made something more from dust would imply that God was less than perfect.

Not only did God have a better material to make woman from, woman was the last creation, and every creation by God had gotten better and more complex.

Sarai And Abram

Now Sarai Abram's wife bare him no children: and she had an handmaid, an Egyptian, whose name was Hagar. And Sarai said unto Abram, Behold now, the Lord hath restrained me from bearing: I pray thee, go in unto my maid; it may be that I may obtain children by her. And Abram hearkened to the voice of Sarai. (Genesis 16:1-2)

Notice, Sarai is telling Abram what to do. Likewise, look at how Jacob is the one being obedient in the following passage.

Leah and Rachel

> *And Reuben went in the days of wheat harvest, and found mandrakes in the field, and brought them unto his mother Leah. Then Rachel said to Leah, Give me, I pray thee, of thy son's mandrakes. And she said unto her, Is it a small matter that thou hast taken my husband? and wouldest thou take away my son's mandrakes also? And Rachel said, Therefore he shall lie with thee to night for thy son's mandrakes. And Jacob came from the field in the evening, and Leah went out to meet him, and said, Thou must come in unto me; for surely I have hired thee with my son's mandrakes. And he lay with her that night.* (Genesis 30:14-16)

Proverbs 31

According to the Bible, the job description of a woman was presented in Chapter 31 of the Book of Proverbs. Let's go through this chapter and see what we can learn about the role of women according to God.

> ¹*The words of king Lemuel, the prophecy that his mother taught him.*
> ²*What, my son? and what, the son of my womb? and what, the son of my vows?*

Most of The Book of Proverbs was written by Solomon. However, notice that this chapter in Proverbs was not written by Solomon. Would

Solomon have been able to tell us about the right way to interact with women? Notice Lemuel was the son of her vows, of her marriage covenant.

> ³*Give not thy strength unto women, nor thy ways to that which destroyeth kings.*

She stated that Lemuel shouldn't invest his time in women (plural), and the reason was given with the rest of the verse in addition to the other ways that destroy kings. Again, not only couldn't have Solomon written this, this chapter was condemning Solomon. Remember, Solomon did the four causes of God's judgment.

> ⁴*It is not for kings, O Lemuel, it is not for kings to drink wine; nor for princes strong drink:*
> ⁵*Lest they drink, and forget the law, and pervert the judgment of any of the afflicted.*
> ⁶*Give strong drink unto him that is ready to perish, and wine unto those that be of heavy hearts.*
> ⁷*Let him drink, and forget his poverty, and remember his misery no more.*
> ⁸*Open thy mouth for the dumb in the cause of all such as are appointed to destruction.*
> ⁹*Open thy mouth, judge righteously, and plead the cause of the poor and needy.*

She stated the job of a king was to remember the law, enforce judgment, and plead the cause of the poor and the needy. Getting involved with women and strong drink would distract the king from his duty. She then stated the value of strong drink was to help ease the pain of those with a heavy heart. We saw that Solomon did not follow this advice.

> ¹⁰ *Who can find a virtuous woman? for her price is far above rubies.*

She stated a worthy woman was extremely valuable. Notice, the introduction of what followed focused on a worthy woman. What was being discussed was the value of a worthy woman. This is the job description of a woman. Notice how the woman was repeatedly stated to be a provider because she is profitable.

> ¹¹ *The heart of her husband doth safely trust in her, so that he shall have no need of spoil.*
> ¹² *She will do him good and not evil all the days of her life.*
> ¹³ *She seeketh wool, and flax, and worketh willingly with her hands.*
> ¹⁴ *She is like the merchants' ships; she bringeth her food from afar.*
> ¹⁵ *She riseth also while it is yet night, and giveth meat to her household, and a portion to her maidens.*
> ¹⁶ *She considereth a field, and buyeth it: with the fruit of her hands she planteth a vineyard.*
> ¹⁷ *She girdeth her loins with strength, and strengtheneth her arms.*
> ¹⁸ *She perceiveth that her merchandise is good: her candle goeth not out by night.*
> ¹⁹ *She layeth her hands to the spindle, and her hands hold the distaff.*
> ²⁰ *She stretcheth out her hand to the poor; yea, she reacheth forth her hands to the needy.*
> ²¹ *She is not afraid of the snow for her household: for all her household are clothed with scarlet.*
> ²² *She maketh herself coverings of tapestry; her clothing is silk and purple.*

23 Her husband is known in the gates, when he sitteth among the elders of the land.

24 She maketh fine linen, and selleth it; and delivereth girdles unto the merchant.

25 Strength and honour are her clothing; and she shall rejoice in time to come.

26 She openeth her mouth with wisdom; and in her tongue is the law of kindness.

27 She looketh well to the ways of her household, and eateth not the bread of idleness.

28 Her children arise up, and call her blessed; her husband also, and he praiseth her.

29 Many daughters have done virtuously, but thou excellest them all.

The intangible qualities are covered. In fact, we have covered the four causes that God uses to bring judgment: pride, fullness of bread, idleness of time, not strengthening the arm of the poor. A worthy woman does the opposite of the four causes of God's judgment.

The job description of a woman involves planning. It involves the long-term aspects of life. Proverbs 31 essentially stated women are the providers because they are responsible for the most profitable aspects of life. She is not involved in any activity that requires physical strength or endurance.

1611 KJV I Esdras

In 1604, King James commissioned a new version of the Bible to be interpreted from the original texts. It was the most intensive translation and involved 47 scholars. In 1611, the King James Version of the

Bible was introduced with additional books between Malachi (last Old Testament book) and Matthew (first New Testament book).

Even though the Puritans continued to prefer the Geneva Bible, they began to reject the books between Malachi and Matthew that were included at that time. In an attempt to get the Puritans to accept the King James Version over the Geneva Bible, these books were removed in 1629. These books, which documented Israel's complete rejection of God, became hidden and obtained the name *Apocrypha*, which means hidden. (Have you ever wondered why there is a 400-year gap between the Old and New Testaments?) One of these books originally included in these earlier Bibles was I Esdras, with chapter 3 telling the following story.

Three Israelites that guard the king (Darius) made a deal that whoever spoke the wisest sentence concerning what was the "strongest" should be declared the winner.

Remember, this was in the 1611 King James Version; however, Strong's Concordance didn't cover the Apocrypha books. The word strongest did not occur in the New Testament, which was written in the same language as the Apocrypha (Greek). However, the word strongest did occur one time in the Old Testament, near the end of the chapter immediately before Proverbs 31.

According to Strong's Concordance it was #1368 gibbor "from the same as 1397; powerful." We saw that powerful is the ability to make things happen.

> Strong's #1397 geber "from 1396; prop. a valiant man or warrior"
> Strong's #1396 gabar "to be strong; by impl. to prevail"

When the three men talked about strongest they were trying to state what ultimately was the most able to make things happen; to prevail. The three answers were wine, the king, and women — but above all things truth bears away the victory.

Notice, we are talking about three people speaking a wise sentence to a king and the answers were: king, wine, and women. Remember Proverbs 31? It began with a mom speaking wisdom (it was even called prophecy) to a king, and the beginning dealt with being a king, wine, and women.

King Darius was not only accepting of this contest, he made a show of it by having each person publicly justify their explanations; to give the *why*! Chapter 3 ended with the first giving his explanation: Wine makes a profitable person unprofitable and a person unaware of an unprofitable situation. He said what Lemuel's mother said in Proverbs 31! In fact, this was the reason she told Lemuel not to drink wine. Wine's strength was argued by how it can make profitable things unprofitable. This was a *resolve* argument.

I Esdras 4 began with the second person giving his reason why he believed the king was the most powerful. The explanation perfectly fits the description of a boss, not a leader. Basically, the king can make anyone do what he wants, and he gets all the benefits. This answer demonstrated a complete misunderstanding of leadership. This was a *solve* argument.

The third person was Zerubbabel, and he gave his reason why he believed women were most powerful. He began by showing women were above wine and the king because women were the cause: they produced life. He even brought up women making garments like in Proverbs 31.

Zerubbabel concluded his explanation about women by hitting all the points. Men look to give of themselves for a woman. They cleave to her, give up the physical, risk their life, make mistakes, and want to please the woman with the physical. Zerubbabel not only confirmed the correct interpretation of cleave, he gave the only answer that resulted in profitability for all.

Zerubbabel concluded his presentation by stating that truth was ultimately stronger than women. He essentially stated that "women" was the answer from man's physical perspective, while truth was the mada answer from a spiritual perspective. Truth was most excellent. Women were most "excellent" and definitely more "excellent" than men, where "excellent" is man's definition that is completely focused on the physical: appearance, power (boss), strength, and wealth. This was a *dissolve* approach.

Notice, Zerubbabel called truth a *she* because it produced life. Basically, Zerubbabel was stating that there was no unprofitability in truth, so it is strongest, most powerful, and most able to make profitable things happen because it was the source of profitability. In the Bible, did profitability always require a woman?

Zerubbabel was declared the winner, and the king said he could be called the king's cousin and ask for anything. What did Zerubbabel ask for?

Zerubbabel asked Darius to keep the vow that was already made: to give the command to build the temple and Jerusalem. Darius kissed Zerubbabel and gave the command to build Jerusalem.

Now you know the event that caused the trigger to God's complex prophecy from Daniel about Him having the ability to bring about His universal will. Why isn't this story in our Bibles?

When we look at the wall, we can see the entire right half is covered by a faint shade representing women!

Since it looks like we have been following a different doctrine relative to women, is it possible we have also been wrong about men? Let's look at men next.

Dr. Joel Swokowski's Commentary

Subject vs Submit

Women are naturally more excellent than men. As a group, men are physically stronger and demonstrate longer endurance than women. Notice, both of these are physical attributes, which we have seen is never profitable in the long term. When it comes to everything else, as a group, women can do what men do, and more! What we have largely been experiencing, in the world and in the church, is men holding women back in the areas that would bring long-term profitability to all of us.

For example, why does a denomination that hinders women from being leaders in church believe it's okay for women to hold roles in music or children's ministry positions? It really comes down to the definition of leadership.

This raises the distinction between the words subjection and submit:

- Subjection is forcing a person to take a lesser position against their will.
- Submit is the willful decision to take a lesser position, which implies they are actually greater.

An example of submit is how a husband ought to act towards his wife, like Christ with the church. He gave Himself for her. The wife ought to take direction from the husband, and everything the husband says ought to benefit the wife, *not* the husband. A wife has every right to ask her husband how what he is asking her to do is for *her* benefit.

If he is a believer, she can further ask if he is hearing from God in what he is directing her to do.

Likewise, we are supposed to submit our will to God (which is always for our long-term benefit). That has been the point of this edition, showing you that people can act in opposition to God's will because we are a first cause and God responds to us. In the moment, our will is superior to God's because God is resting; He has ceased from His occupation of being the first cause. God moves in response to spiritual value (justice), either positive ($100 bills) or negative (pennies).

If you're struggling with the phrase, "our will is superior to God's," keep in mind the reason, "because God is resting." Also, our will being superior to God gives us the ability to submit to Him and His will. Furthermore, ask yourself, whose will is happening more often in this world, God's or man's?

We've already learned that a leader facilitates the purpose and progress of another, and a boss facilitates their own purpose and progress at the expense of another. This makes being a servant the effect of being a true leader.

When it comes to women, what about Deborah the judge in Judges 4? She was God's leader to His people! Or Phoebe the deaconess from Romans 16? Or Huldah (prophetess) who helped Josiah and Israel in II Kings 22?

What is the church guarding against? Does the church want more leaders, or is it satisfied with people in a position of authority gaining at the expense of others? There can only be one boss, but everyone can be a leader. Bosses don't like to be bossed, but leaders like being led.

Women can and should be leaders, if they want to. This is easy to see when you have God's definition of leadership: one who facilitates the purpose and progress of others!

What is the goal of feminism? Is it for women to be treated equally with men? Doesn't this actually put women in a lower position than they should be? If women are superior to men in the areas that are profitable, then making them equal to men would result in limiting their ability to be profitable by lowering them to the level of men.

Remember, this entire work is attempting to present the greatest system: God's will. In order to appreciate the system (whole), you have to have all the parts. This means a person shouldn't be making conclusions until they have all the parts.

For example, remember when we covered God's objective measure for virginity? I've found people are so inured with how far we have strayed from God's will, they will attack this measure forgetting they are attacking God! Why would God make women responsible for carrying the burden for virginity when we believe it is not an objective measure?

When people tell me that they learned in health class that a woman who is a virgin may not bleed on her wedding day, I ask them, "What was the reason given?" If they aren't hypocrites (from quoting an effect without retaining the cause), they will say, "A woman's hymen can tear from strenuous physical activity (e.g., gymnastics), trauma (e.g., bicycle accident), or using a tampon."

How many of these things were covered in Proverbs 31? Did God intend for females to focus on physical activity, distracting them from developing their ability in the profitable areas? Unfortunately, we have strayed so far from God's will, and we don't participate in systems thinking like Jesus and the Apostle Paul.

Man-made tradition argues for the way it is. God argues for the way it ought to be. Which do you argue for?

CHAPTER 14

Men

THE OXFORD LANGUAGES dictionary gives the definition for masculinity as "qualities or attributes regarded as characteristic of men or boys." This is not a definition; rather, it is a description containing effects. If I were to summarize the way the word masculinity is used in our society, I would arrive at a causal definition of "wildness."

This definition of masculinity society uses today was created during the Industrial Revolution. Before the Industrial Revolution, home and work were in the same location. This is obvious with a farm, but even the downtown stores were both: the front was the family business, and the building's back (or upstairs) was the home. Before the Industrial Revolution, men and women both worked, and both raised the kids.

It was only once home and work were divided that men worked and women stayed at home to raise the kids. At that point, society's definition for masculinity became: "strength, dominance, independence, and ambition." These are the traits it took to rise to the top of the corporate pyramid. Eventually, this could be summed up as "wildness." Today, people can't define masculinity in a non-contradictory way, so they speak of toxic masculinity because they can only speak about what they don't want.

It may interest you to know that all the parenting books written before the Industrial Revolution were to the father (see *Mothers and Such: Views of American Women and Why They Changed* by Maxine L. Margolis). Look at the statistics for single fathers vs. single mothers. Kids are much worse off if they are raised by a single mother than a single father. For example, according to Daniel Amneus, Ph.D. (*The Garbage Generation*), research shows children are twenty times more likely to end up in jail being raised by a single mom vs. a single dad. It would seem the quickest way to ruin society is for the fathers to be removed from the home and for those "men" to be encouraged to act like animals.

Society's definition for masculinity before the Industrial Revolution was "the ability to put the interests of others above your own" (*The Creation of the American Republic, 1776-1787* by Gordon S. Wood). This is really tough. That's why it takes a really tough person to put the interests of their wife and kids above their own. How many men fit this definition of masculinity?

What does the Bible give as the definition of man?

Definition of Man

There is a verse in the Bible that gave the definition of *man*:

> *Be of good courage, and let us play the men for our people, and for the cities of our God: and the Lord do that which seemeth him good.* (II Samuel 10:12)

Joab put the men in line for battle and then stated they should *play the men for our people*, which meant focusing on doing the causes. Someone needed to address the issues, and courage was required, so what else would this be other than addressing the issues/causes? What about the effects? Joab said, *the Lord do that which seemeth Him good*. Joab left the effects to God. He didn't focus on them at all.

Notice, our society needs people who do the causes even when they know the effects are unprofitable. Why? Because if they don't do the causes, society will become catastrophically unprofitable.

Think about war, a burning building, or a child that has fallen down a well. When each of these situations is handled as well as possible, is the result profitable? Are the soldiers better off after the war than they were before the war? Is the house that caught fire better than before it caught fire? Is the kid healthier than before they fell down the well? No. No. No.

However, if someone doesn't handle these types of situations, there will be more deaths in the war, the house will be completely destroyed, and the child will die. All of these situations become catastrophically unprofitable. Can you see how "wildness" can be confused with men handling these situations? While these circumstances require a man to be brave, he shouldn't be wild and reckless.

A healthy community requires people who will do the causes even when the effect is going to be unprofitable; otherwise, things will be much worse. We need people who will do the causes knowing they will not personally benefit from the effects, whether the effects are profitable or not. These people are known as men.

> **Man:** a male that is intentionally able to focus on (do) the causes regardless of the effects.
> **Boy:** a male that is focused on effects.

This explains something I've noticed for decades: every time a male does damage, it is because he is focused on the effects. Worse, he tries to make effects into causes.

Notice, if the male is focused solely on the causes then he cannot completely control the effects. So the first effect of being a man is that he is humble. He can only stay focused on the causes if he humbles himself and allows God's Nature to flow through him.

We asked a question in the previous chapter and now we have the answer. The best a group of men can achieve is not unprofitable. There are no examples in the Bible of a group composed solely of males being profitable in the long term. To answer the question in the previous chapter: Yes, it does take a woman to make a situation profitable in the long term.

In *Modeling God*, we saw that even The Trinity is not unprofitable. How could The Trinity interact with each other in a manner that results in gain? Everything The Trinity created has returned less value than what it took to create it. To think that we have returned more value to The Trinity than what They gave to us is to make us superior to The Trinity.

In the last chapter, we saw how man's will can be superior to God's will because God is resting. This is different from thinking we can return more value than The Trinity. Be careful.

Notice, men and women ought to both do the causes (grace, faith, etc.); however, women tend to naturally do the causes, while males do not. It's just one of the behaviors that females don't seem to understand about males.

Jesus is the ultimate man. Jesus stepped into a situation that needed someone to be completely about the causes knowing that the result was going to be unprofitable. However, Jesus also knew that without Him, the result would be catastrophically unprofitable.

Take a moment and realize what it would be like to know ahead of time that all the effort you will give for three and a half years will still end up in your death and your followers eventually being murdered, and that is the best possible result! Remember, even though Jesus provided the value for people to achieve salvation, many more people are going to end up in the lake of fire.

Next time you read a gospel account of Jesus, notice how He doesn't get excited about effects, both good and bad. Notice how He constantly reinforces the causes. In fact, He saw every cause as spiritual/intangible, not physical/tangible.

In eternity, Jesus will continue to always and completely do the causes. We, as the Bride, will learn to make profitable effects from the causes He gives to us. Even when we don't do the effects, the overall result will be not unprofitable.

This eternal plan can never become less! This is how God can be sure His eternal plan will continually result in more. However, if Jesus had been the Bride and the church came together to be the Groom, it would be possible for eternity to become something less.

Much like the way to world is being run now, we could take the first step as the Groom and make things unprofitable, and then Jesus as the Bride would have to spend a lot of effort trying to get things back to not unprofitable, but it would always end up being unprofitable in the long term. In the same way, one spouse can always find a way to spend more money than the other spouse earns, a Groom made up of humans would eventually be able to find a way to make things catastrophically unprofitable.

I believe these two man-made doctrines, leadership and the role of the sexes, are the main causes that make the word and will of God of none effect, that is, catastrophically unprofitable. These man-made doctrines are the causes that make church and marriage unable to efficiently generate the spiritual value to bring about God's will on earth as it is in heaven.

I believe God's definition of leader results in people becoming more able to do God's will over time. In fact, I believe the man-made, contradictory, traditional definition of leader (boss) actually results in us getting increasingly further away from accomplishing God's will!

Think about the traditional definition of leader (boss). It is not possible for all of us to facilitate our own purpose and progress at the expense of others.

Leaders operate according to causality. They provide the cause in order to get the desired effect.

Bosses operate according to the opposite of causality. They attempt to get the effect without providing a reason or a cause.

We found that we are all eventually unprofitable discussing leadership because too many people are actually describing bosses. We saw that Solomon began with mada focused on causes. However, he ended up focusing on effects to the point that he stated everything was pointless (unprofitable; not getting desired effects) at the end of his life, and the four causes of God's judgment led to Solomon's destruction.

The reason pastors were men in the Bible is that they ought to be about the causes regardless of the effects. Teaching the apostles' doctrine, fellowship, breaking bread, and praying are all causes; things that can be intentionally done regardless of the effects. The church ought to be like a woman, able to make profitable effects from the causes it is given.

Sarai, along with Leah and Rachel, showed how the relationship between men and women in Genesis was the opposite of what we have today. When did this flip?

The Flip

I Samuel 8 told the story of the Israelites saying they didn't want God's way of leadership, which was through a judge. The first judge was Moses and the judge during the time of the story was Samuel. Instead, the people wanted a king (boss) like the other nations. Here was how Samuel and God felt about it.

> *But the thing displeased Samuel, when they said, Give us a king to judge us. And Samuel prayed unto the Lord. And the Lord said unto Samuel, Hearken unto the voice of the people in all that they say unto thee: for they have not rejected thee, but they have rejected me, that I should not reign over them. According to all the works which they have done since the day that I brought them up out of Egypt even unto this day, wherewith they have forsaken me, and served other gods, so do they also unto thee. Now therefore hearken unto their voice: howbeit yet protest solemnly unto them, and shew them the manner of the king that shall reign over them. (I Samuel 8:6-9)*

God considered this request the same as rejecting Him. God put Samuel in His (God's) place with respect to prayer. He told Samuel to listen to their request, protest solemnly, and then show them the implications for what they want. Likewise, I believe God speaks to us after we make our request during prayer, and then we can decide if we still want what we asked for or if we will listen to His counsel. Prayer is a council meeting! This chapter seems to be central to understanding God's will! Here was what Samuel said.

> *And Samuel told all the words of the Lord unto the people that asked of him a king. And he said, This will be the manner of the king that shall reign over you: He will take your sons, and appoint them for himself, for his chariots, and to be his horsemen; and some shall run before his chariots. And he will appoint him captains over thousands, and captains over fifties; and will set them to ear his ground, and to reap his harvest, and to make his instruments of war, and instruments of his chariots. And he will take your daughters to be confectionaries, and to be cooks, and to be bakers. And he will take your fields, and your vineyards, and your oliveyards, even the best of them, and give them to his servants. And he will take the tenth of your seed, and of your vineyards, and give to his officers, and to his servants. And he will take your menservants, and your maidservants, and your goodliest young men, and your asses, and put them to his work. He will take the tenth of your sheep: and ye shall be his servants. And ye shall cry out in that day because of your king which ye shall have chosen you; and the Lord will not hear you in that day.* (I Samuel 8:10-18)

Samuel explained the effect of their request was that they would end up with a boss. However, the main implication was what wasn't directly stated: you will value males over females because tangible stuff was the goal. Everything God had set up would be flipped around. It was from this point that all the misogynistic claims people make concerning the Bible occurred, but it was not God's will.

I Samuel 8 showed how everything flipped away from God's will and the chapter covered prayer (God's will), leadership, and the role of the sexes! Let's look at a story that occurred prior to I Samuel 8 to see how men and women ought to interact.

Ruth

The Book of Ruth told a story that occurred approximately 160 years before Saul became king. A Moabite woman, (Naomi) married an Israelite (Elimelech). After they had two sons while living in Moab, Elimelech died. The sons both married Moabite women and after ten years, the sons both died. When Naomi decided to return to her homeland of Bethlehem in Israel, one of the two daughters-in-law (Ruth) wanted to accompany Naomi. Despite Naomi trying to convince her otherwise, Ruth returned with Naomi.

In Bethlehem, the women gleaned in a field owned by Boaz and he became attracted to Ruth. Boaz wanted to marry Ruth; however, the law stated that there was one man who was a nearer kinsman to Ruth and he had the first right to marry Ruth. Let's pick up the story in Ruth chapter 4 when Boaz confronted the near kinsman in front of the elders.

> *And he said unto the kinsman, Naomi, that is come again out of the country of Moab, selleth a parcel of land, which was our brother Elimelech's: And I thought to advertise thee, saying, Buy it before the inhabitants, and before the elders of my people. If*

thou wilt redeem it, redeem it: but if thou wilt not redeem it, then tell me, that I may know: for there is none to redeem it beside thee; and I am after thee. And he said, I will redeem it. (Ruth 4:3-4)

When Boaz asked the near kinsman if he wanted to own the land that belonged to Elimelech, the near kinsman stated he wanted it. Now, watch what happened next.

Then said Boaz, What day thou buyest the field of the hand of Naomi, thou must buy it also of Ruth the Moabitess, the wife of the dead, to raise up the name of the dead upon his inheritance. And the kinsman said, I cannot redeem it for myself, lest I mar mine own inheritance: redeem thou my right to thyself; for I cannot redeem it. (Ruth 4:5-6)

What just happened? The near kinsman wanted the land when Ruth wasn't involved. However, once he found out he would have to marry Ruth, he said he didn't want the land. Remember, he can be married to more than one wife. What man would turn down the opportunity to have more land and another wife?

Again, this story took place before Saul became king. The near kinsman knew he would have to husband Ruth, which was a responsibility. He would have to grow her into what God made her to be by giving her as much attention as he gave his first wife and her kids, or he could be put away. He knew he would have to have kids with Ruth, and those kids would be raised up in the name of the dead. This would take attention away from his own inheritance by giving his first wife and her kids less attention, even though he would have more land.

Proverbs 31

Take a big step back and summarize the job description we read in Proverbs 31 for women. What didn't it cover? It didn't cover working by the sweat of one's brow, fighting, going to war, risking her life, handling confrontation and crisis, etc. All the things that require physical strength and endurance.

We saw the job description of a woman involved planning. It involved the long-term aspects of life. Proverbs 31 involved the most profitable aspects of life. Women were actually the providers! This meant men were made to handle crisis, conflict, physical exertion; the short-term aspects. Men were made to put out fires at a moment's notice. Now we can see the definitions for woman and femininity.

> **Woman:** a female that is intentionally able to create profitable effects from causes she is given.

We saw the definition of masculinity, but what about femininity? The definition of *femininity* is "the ability to nurture others," where nurture others means to provide necessities for life. Remember, life is the ability to repair; profitability. We can even see this as the ability to provide the necessities of life. This also gives us the information needed for yet another definition.

> **Girl:** a female who attempts to create profitable effects from causes she gives herself.

In order for a woman to provide through creation, she needs safety, which is the responsibility of a man; protection. As a group, males are physically stronger and able to demonstrate greater endurance than females. If males use these abilities to provide safety, then they are being leaders and men. If males use these abilities to physically intimidate females, then they are being bosses and boys. There's another huge variable we need to account for to completely understand how men and women ought to interact in order to be profitable.

Males tend to only be able to have one thought at a time. God made male brains in this fashion. Females tend to be able to have up to five thoughts at the same time in a healthy manner. God made female brains in this fashion.

Notice, it's not that a woman can't do what a man can do. The question is why would you want her to waste her ability being about the causes regardless of the effects? For example, I can use a smartphone as a coaster for my drink, but why would I, especially if a coaster is available?

Remember, everyone wants to be happy. The issue is how people go about being happy. What is the man's God-given desire for happiness? Men want to never get tired of having sex with the same woman. How do men go about achieving this desire?

Their strategy is to have sex with the most beautiful woman. We have seen that God made everyone's brain to build a tolerance to external, tangible stimuli. Consequently, this strategy focusing on the physical not only runs down, the male gets tired of having sex with her. Show me the most beautiful woman in the world, and I will show you a guy who got tired of having sex with her.

The only way for the male to achieve this goal in a healthy, profitable, and generative manner is for the man to help the woman continually grow into becoming everything God intended from her uniqueness, which is husbanding! If she continually grows, he will continually be having sex with a more excellent woman. We have seen the way to help anyone continually grow and strengthen their marriage is through sharing, and repair, after abuse.

The opposite of this is a woman husbanding a man, which was not included in her Proverbs 31 job description, while the men hanging out at the city gate were husbanding each other (Proverbs 31:23). Now we see that husbanding is something the woman (and all of us) need in order to grow.

We have seen God's strategy was for a man to focus on helping his wife be profitable and achieve success. What would the opposite strategy look like? What would it look like for the woman to focus on helping her husband achieve success? Who is the only woman that was completely about each male? Mommy. Wives, do you want to have sex with your son? Husbands, do you want to have sex with your mom?

Recently, we've had more than one example of extremely rich men having sex with other men's wives. Do you think these wealthy men were happy? Once these men became wealthy and powerful, their wives gave up their careers and focused on helping their husband achieve success. Now that they were having sex with "mommy," it isn't too difficult to see why they thought they needed to have sex with another man's wife in order to be happy.

What is your strategy for happiness?

Summary

Now we can see the entire left half is covered by a faint shade representing men! We are now far enough away from the wall to see how leadership and the role of the sexes relate to church and marriage, which is God's will!

Throughout the Old Testament, God used marriage and harlotry to illustrate His covenant with people. The New Testament ended with an illustration of a Husband and a Wife in their proper roles and the marriage lasts for eternity. Jesus, the ultimate man, will spend eternity husbanding, that is, developing the Bride into being everything She was made to be, and it was described as the joy set before Him for which He came down from Heaven to endure the cross (Hebrews 12:2).

Why don't males today see husbanding as the role that is going to bring the ultimate fulfillment?

God gave us marriage as a blessing in order to bring about God's will.

God's ultimate plan, the meaning of life, is church and marriage. God is able to achieve His plan for continually experiencing more because of His last and best creation: the Bride.

The Trinity interacting with each other through the Bride is the source of eternal profitability.

Dr. Joel Swokowski's Commentary

Lenhart shared the non-contradictory definition of *man* is "a male that is intentionally able to focus on (do) the causes regardless of the effects." Although we've learned what causes and effects are, a simpler (and frankly less technical) way to see this is "a man does the right thing regardless of what he gets out of it. He does the right thing because it is the right thing!"

Lenhart stated the following:

Think about war, a burning building, or a child that has fallen down a well. When each of these situations is handled as well as possible, is the result profitable? Are the soldiers better off after the war than they were before the war? Is the house that caught fire better than before it caught fire? Is the kid healthier than before they fell down the well? No. No. No.

It's clear we need men to deal with the fires being put out. This brings up another great example of what it means to be a man. Imagine there's a building on fire and two males run into the building to save a family.

One male runs into the building because he sees the camera crew there and he wants to be on the news. He also knows this will likely result in him getting a promotion.

One male runs into the building…because it's on fire.

Which is the man? Which is the boy?

Basic Husbanding

There are two basic steps to husbanding.

1. Spend ten to fifteen minutes each day helping her remove two to three thoughts.
    - Here, the husbandman checks in on how the husbandry is doing. "What are you thinking? Feeling? How was your day? etc." He has learned the unique way that facilitates the husbandry to share what is on their mind.

- It's good practice to ask a question like, "What else?" after they share a thought in order to ensure you're removing enough thoughts to keep their thought process clean.
- The key is to not tell the husbandry how to address the issue or solve the issue. Speak with her until she has filed it away herself. It turns out that females tend to be able to handle up to five thoughts at the same time in a healthy manner. It's when they add that sixth thought that they crash. If a man helps her remove two to three thoughts each day, she won't crash. That begs the question, "When she crashes, whose fault is it?"

2. Answer every issue with, "How would you handle it if you were more yourself?"
    - The first step is really meant to avoid unprofitability. We are avoiding the crash. At some point, she will have a complex situation that she can't file away, so she is going to look for the husband to solve it. The whole point of husbanding is to help the husbandry become what (who) they were created to be. The way to address this is to ask her how she would handle it if she was more herself.
    - I may have a way to handle the situation, but it's my way, according to my uniqueness. The husbandry has their own unique purpose and ought to learn to handle those issues according to that purpose.
    - Notice, this implies that the husband and

husbandry have identified and agreed to the purpose of the husbandry.

The Trinity

Lenhart wrote what followed above; however, I'm quoting it here to help you take this point in carefully. Keep in mind what Lenhart is saying here and recognize that he is not saying God is less than perfect. In fact, it is the Trinity's perfectness that results in the effects Lenhart taught:

In Modeling God, *we saw that even The Trinity is not unprofitable. How could The Trinity interact with each other in a manner that results in gain? Everything The Trinity created has returned less value than what it took to create it. To think that we have returned more value to The Trinity than what They gave to us is to make us superior to The Trinity.*

In the last chapter, we saw how man's will can be superior to God's will because God is resting. This is different from thinking we can return more value than The Trinity. Be careful.

In my commentary from the "2. God the Creator" chapter of *Modeling God*, I wrote the following:

A question to really expose a person's view and definition for God is, "Can God grow?" If you say "yes," then you are stating God is less than perfect. If you say "no," then you are saying there is something God can't do. God being always completely righteous and always completely just is the only explanation that proves God is the greatest being in existence. What's better than always completely righteous and always completely just?

This point was made to emphasize the fact that God being always and completely righteous and always and completely just was the highest of principles. So high, God cannot grow. This same point proves that the Trinity, in and of themselves, cannot become more than what they ARE (Jesus being on earth is the one exception due to Him being both fully God and fully man while in a human body). However, God knew that the only way He could experience happiness in eternity is by the creation of the Bride. We will discuss the details of God's eternal plan in the third and final edition.

More Prayer Implications

In my commentary after the "5. Prayer" chapter in *Modeling God*, I wrote the following:

Go ahead and look back at the Prayer Model. I want to give you a way that might make it easier for you to remember what your responsibility is as it relates to requesting a value from our Lord. I call them the Four R's of Prayer:

1. *Recognize God*
2. *Reinforce your Faith.*
3. *Reference Justice.*
4. *Request and let it go.*

What we learn from the I Samuel 8 story is that God is trying to get our attention while we pray. Prayer is "two-way" in that it ought to be seen as a back-and-forth conversation with God through the entire prayer, not just that I go through the entire prayer and then wait for His response. God wants to help us pray in a way that is most effective, meaning we may end our conversation with God without making a request at all! That's what ought to have happened with the Israelites. They ought to have heard God's rebuttal and changed their minds about their request for a king.

This brings up what I would call the 5th R of Prayer: God's **Rebuttal**. Throughout the prayer, we ought to keep an ear to God for any information He may give us that would help us understand the implications of our request. God is trying to talk to us all the time, even when we are praying to Him.

God is part of the prayer process, more than just deciding if He'll grant your request. We know our responsibility in how to pray. Here's what God is doing:

1. Listening
2. Protesting (Rebuttal)
3. Checking for your final answer (Are you sure?)

Part of our responsibility in praying is to keep an ear to God while He does His part. If we ignore this responsibility, He will take whatever our request is as our final answer, regardless of how detrimental that request may be! (Think of Manasseh!!!)

CHAPTER 15

Bringing About God's Will

IF EVERYONE HAD chosen to be saved and sanctified, allowing God to do His will through them, God would already have enough spiritual value to bring about His eternal plan, and no one would have ended up in the lake of fire.

Instead, a large majority of people are indirectly facilitating God's will at their own expense. They are doing their will, which allows God to accumulate spiritual value through justice by giving the truth with love and forgiving in order to get spiritual value to bring about His eternal plan; however, this delays God's ability to begin His eternal plan and results in more people spending eternity in the lake of fire.

This book has shown how the main cause of this delay is our opposition to God's doctrines of leadership and the role of the sexes, which hinders the means God provided to bring about His will, church and marriage, especially when we consider God's eternal plan is church and marriage.

I believe there is one group of people who are delaying God's will worse than anyone. We have stated the same principles that it takes to make money are the same principles it takes to create spiritual value. This means the people who have made the most money ought to be the ones who lead all of us in making the most spiritual value. This is especially true when we learn about the "prosperity paradox": the more money a person makes, the more unhappy they are. Clearly, these intelligent people ought to be the first to realize that money won't make them happy and, therefore, should pursue spiritual value.

Instead, these people not only focus more on making money, they distract the rest of the population away from accumulating spiritual value by trying to convince everyone money is making them happy. Not all people with tangible riches are evil; however, the most evil people in the world are the ones with tangible riches.

Isn't this what we saw with Solomon? How many people think Solomon was a success? How many people realize he finished his life unhappy and unable to facilitate God's will? Solomon opposed God's will by turning his heart away from God and leaving the idols erected to other gods in the high places leading the population in opposing God's will to the point the nation of Israel split immediately after Solomon's death. Even after all that, the majority of people today are distracted and deceived into trying to replicate Solomon's life!

Ultimately, a person's choice of whose will they are pursuing is proven by whether they focus on the quantitative and tangible (more stuff) or the qualitative and intangible (spiritual value). It turns out, determining who is directly helping or hurting God's will is easier to spot than what we may have realized.

Now that we understand how the biggest system works, we know the first step in addressing any system we encounter: give up control! When our first step is to make immediate progress, I call this taking a *forward* step. Taking control is a forward step; it is analysis. If the issue involves people, the forward step will cause three more problems. When our first step is to the bigger system, I call this taking a *backward* step. Giving up control is a backward step.

> *Delight thyself also in the Lord; and he shall give thee the desires of thine heart. Commit thy way unto the Lord; trust also in him; and he shall bring it to pass. And he shall bring forth thy righteousness as the light, and thy judgment as the noonday.* (Psalm 37:4-6)

What are the desires of your heart? The word desires came from H4862 misala and meant requests. These are the tangible activities that are most inline with who you are created to be. The verse shows us these desires will come from God as an effect of focusing on God's will! Basically, your desires are on the path of God's will! Here's Jesus stating the same thing:

> *But seek ye first the kingdom of God, and his righteousness; and all these things shall be added unto you. Take therefore no thought for the morrow: for the morrow shall take thought for the things itself. Sufficient unto the day is the evil thereof.* (Matthew 6:33-34)

We have seen God is Right-Right by taking a backward step and only moving in response to justice. Likewise, this is the way we are able to dissolve an issue in any system. The ultimate way is to give up control and take direction by the Holy Spirit.

The rest of this chapter is meant to give you objective steps in order to help you not only be in God's will but also facilitate God's will by giving up control to the Holy Spirit.

Facilitating God's Will

1. Remember, the Holy Spirit can only move in response to justice.

Notice, justice is two-way. When you are unjust, the Holy Spirit can move to equal out justice against you! Guess what! We are already experts at getting the Holy Spirit to move half of the time! Yay!

This step is meant to help you make sure you aren't hindering God's will when you want the Holy Spirit to move to equal out the injustices done to you.

The key point to realize is the Holy Spirit is completely aware of injustices done against you. In fact, the Holy Spirit knew about them before you did! Any time you spend wondering why God doesn't see the injustice of your situation is wasted time. It is actually you delaying the result you want to see!

Here is a scripture I used to have over my mirror so I saw it every morning:

> *For the eyes of the Lord run to and fro throughout the whole earth, to shew himself strong in the behalf of them whose heart is perfect toward him.* (II Chronicles 16:9a)

God wants to move on your behalf! The next point is another way to stop frustrating God!

2. Become an expert at resolving disagreements with covenant partners.

Likewise, ye husbands, dwell with them according to knowledge, giving honour unto the wife, as unto the weaker vessel, and as being heirs together of the grace of life; that your prayers be not hindered. (I Peter 3:7)

Therefore if thou bring thy gift to the altar, and there rememberest that thy brother hath ought against thee; Leave there thy gift before the altar, and go thy way; first be reconciled to thy brother, and then come and offer thy gift. (Matthew 5:23-24)

Whether it is marriage or church, do you know how to resolve disagreements with the person? If you don't, your prayers will be hindered until you get back into agreement. The next step covers how to resolve disagreements with unbelievers.

3. Forgive

Remember, you are connected to everyone in the world (believers and unbelievers) through the Holy Spirit. When you forgive unbelievers, the Holy Spirit can justly work on them from the outside. All the time you spend *not* forgiving unbelievers is wasted time. The ultimate in forgiveness is looking to help the people who have been unjust to you, like Jesus when He forgave the people who were killing Him *while* they were killing Him!

Worse, you are limited in being used according to God's will for God's plan due to your inability to forgive. God cannot put you in a difficult situation and work through you to dissolve it if He can't trust you *not* to work out your own justice!

We have seen how truth begins in the opposite direction. This means that sometimes God's will can look so wrong to us that we trigger. We have an immediate negative response to the point we attack a person (like Pharaoh) or flee the situation (like Jonah). If this occurs while we are in God's good will, not only could it result in us missing out on fulfillment from doing our God-given purpose, we could take ourselves out of God's will.

One of the questions I repeatedly get is: "What if I'm being blamed for something I didn't do?" First of all, this is an opportunity to gain spiritual value if you respond to fixing it as if it were your fault while having a good attitude. Second, what is wrong with intentionally taking the blame for something you didn't do? Take a moment and answer that question before continuing.

Taking the blame for something you didn't do is what Jesus did that resulted in your salvation. The foundational principle of Christianity is bearing the pain of others. Wouldn't a person who had God flowing through them naturally take the blame for something they didn't do? Isn't this the best way to gain spiritual value while also bringing judgment on the person to propel them towards repentance? I think we call this dissolve.

4. Become an expert on doctrine, especially the doctrine of prayer

Prayer is the main way we operate in God's will! A lot of doctrine needs to be understood in order to be an expert on prayer. In the commentary of *Modeling God*, Joel illustrated the gulf between experts and novices through a story looking at the approach two different teenagers use to ask their parents for $50. The disrespectful teenager is emblematic of how the overwhelming majority of people pray to God. The second teenager is how all of us naturally approach authorities in our lives when we truly want something. Joel asked this question: *Why would a person treat the God of the Universe in any manner less than the way the second teenager treated their parents?*

5. State your will

I believe the deep down desire of your heart is God-given, so the first step is to state it out loud. Don't hold back. When people struggle, I help them by asking them, "If you had a magic wand, what would you wish for?"

If you don't state your will, you aren't going *hot* or *cold*, and Jesus said He would spit the church of Laodicea out of His mouth for the same decision. It doesn't matter what your will is; what matters is that you state it! Dr. Russell Ackoff considered this the first step in the dissolve process: Tell me what you want in an ideal world. In this case, you ought to be stating what you believe is God's will.

God cannot move for you if you don't state out loud or write down what you want.

All the time you spend resisting stating your will is wasted time.

6. Give up Control; Faith and Humility

Since you can't *do* a *don't*, giving up control is the most difficult step because we have to have something we can go towards. Of all the steps, this is the most difficult, so please take your time with the following mini-process.

A. **State specifically the ideal way you want your desire to happen.**

The deep-down desire of your heart is given to you by God, so it is not wrong. It is God's individual plan for you. However, how you want this to come about is *your* will. The key is to realize your will is not God's will.

The reason people are frustrated with God's plan for their life is they think their plan and their will are either both right or both wrong. Your plan is right because it is God-given. Your will is wrong because you don't have all the information and your way of bringing about your plan would end up with something less in the long term. Now you can do the most important step.

B. **State that it will never happen your way.**

I like to see this by considering a spectrum. On one end is the easiest and most amazing way your plan could occur. On the other end is the hardest and most arduous way.

Imagine some of the steps on this spectrum of ways God could bring this about going from the easiest possible way to the most difficult. Stating the easiest and most amazing way it could come about places you on the end of this spectrum. This allows you to see the rest of the spectrum at one glance, so you will quickly recognize when God brings the opportunity.

Stating anything less than the easiest way will cause you to be away from the two ends of the spectrum. You will have to keep looking toward both ends of the spectrum to see how God is going to do it.

Also, stating your will as something more difficult than the easiest way opens up the possibility His will is easier than what you expected. This may also result in you missing the opportunity. Besides, I think everyone has faith that the easiest and most amazing way is not going to happen.

In order for the Holy Spirit to move justly, you have to give up control over the effects. That's why the ultimate way to end every prayer is the same way Jesus ended His prayer: "...*nevertheless, not my will, but thine, be done*" (Luke 22:42).

This makes the first part of the prayer a statement of your will on how much of your spiritual value you want to budget based on your request. If you were God, then your prayer would be the perfect way to bring this about, both according to efficiency and effectiveness. You are not God. So, your giving up control over the effects is you saying you are not God.

Effective: Do you believe God can bring about a better final result than what you asked for with the spiritual value you budgeted?

Efficient: Do you believe God can bring about the same result as what you want by using less of your spiritual value?

God has all the information that exists. God is infinitely more efficient and effective than you. This means the simplest version of prayer is two steps:

1. State what you want and how you want it to come about. (Your plan and will)
2. State that you know it will never come about the way you stated. (Not your will)

We saw with I Samuel 8, God does respond *during your prayer* with His will. If you don't have enough faith to hear Him during the prayer, ending the prayer with "…nevertheless, Your will be done" is your way of agreeing with Him regardless of whether you heard what He said or not.

C. **Identify everything you need to do that is hindering the Holy Spirit.**

The Holy Spirit is going to bring about God's will for God's plan for your life; however, there are things that can hinder the Holy Spirit from bringing about the plan. These are hurdles. For example, if you want a new job, you would state your ideal job, how you would *want* to get it, and then that it will never happen that way.

Next, you would decide what are the things you need to do so you don't hinder the Holy Spirit. This could include: writing out the characteristics of the job, writing out your salary requirements, and having a resume you can send out immediately. After all, the Holy Spirit isn't going to bring you a job opportunity only for you to not be able to respond because you don't recognize it or have a resume put together.

If you have trouble coming up with these hurdles, take a moment to pretend it is your job to help someone else get this same dream. What would you tell them they need to do for you so that you could accomplish it?

D. Do the hurdles.

The Holy Spirit is going to bring about this plan if you give up control. Focusing on the hurdles results in you giving up control over the drivers, the aspects you want that are the responsibility of the Holy Spirit.

E. Determine possible reasons God would want you to stay in your current situation.

We are assuming that you know all the hurdles preventing the Holy Spirit from moving. However, even after you do all this, there could be a reason God wants you to remain in your current situation that is not God's individual plan. There are really only three reasons:

1. **Timing:** The circumstances aren't ripe. Notice, this is quantitative. You can be too early (as well as too late) to the opportunity and it wouldn't be good.
2. **Ability:** You haven't become strong enough at an ability you need to have for the next opportunity. This is also quantitative. Yes, you may have shown that ability, but have you mastered it enough to do it under the greater tension you will experience?
3. **Value:** You don't have enough spiritual value. Actually, this one is *always* in play. When people understand this concept, they ask me how much

spiritual value they ought to be trying to get. My answer is, "More…always more." You can't have enough spiritual value.

F. Humility: Consider you are wrong.

Give God the opportunity to correct you. It doesn't matter if you are wrong. It matters your response to being wrong. This was the point of fleeces, as seen with Gideon in Judges 6 and Jonathan in I Samuel 14:8-10.

While I've read a lot of information on fleeces, I rarely read about the ultimate purpose of a fleece. The goal of a fleece is to determine the truth, which is God's will. If you are doing a fleece for any reason other than knowing the truth of God's will, then you are opposing God's will.

In order to allow God to show you His will, the fleece has to be something you are not in control of. In fact, when people tell me about a sudden benefit they have received, I immediately ask, "Did you make that happen?" If not, then I encourage them to immediately give the credit to God.

There are two ways I help people create a fleece:

1. If God wanted you to change your current situation, what would be the proof? (Answer)
   Then, give up control by continuing to address the hurdles and see if the fleece happens.

2. If God wanted you to stay in your current situation, what would be the reason? (Answer)

Then, while you are waiting to see if God is going to change your situation, give up control over the circumstances by working on improving at the reason you would be staying in your current situation.

It is important that we give up control so the Holy Spirit can bring about God's will, and we know for sure that we didn't make any of it happen. The mini-process within this step ought to help you focus on what to *do* so you *don't* take control and can grow in faith.

7. Be a Leader instead of an Enabler

If you are able to do the first six steps, then you have something to give to others, which qualifies you to be a leader. How do you help others with God's will?

God's will can be painful. I'm realizing there are a lot of people who equate "being in God's will" with "no pain," to the point they will even state, "God would never do something that causes me pain." Unfortunately, this is almost everyone's belief when it comes to helping other people!

Does this pattern look familiar?

A. A person you know is doing good and they don't want to interact with you.
B. The person experiences pain and now wants to interact with you.
C. You resolve the pain as quickly as possible.
D. The person is doing good enough and they don't have to interact with you any more.

E. You get frustrated they didn't appreciate your help.
F. They get worse and are on their way to hitting bottom.
G. They reach out to you for help.

One of the first questions I ask when people want advice on how to help someone else is: "Has this person reached out to *anyone* for help or have they received *any* help?" I ask this because I'm trying to find out if the person is at B or F because I want to help you do something other than C or be ready for G.

Remember:
- We are humans.
- We are intangible minds and souls operating through a tangible brain in a quest to release pleasure chemicals.
- We are unique; however, the response is the same for all of us: embrace truth and have community.

Resolving the pain as quickly as possible is enabling if the person doesn't actively grow by embracing truth and plugging into a community. Before I continue, let me pause to say a couple of things about community. The community in the meaning of life is church; however, family is also a community.

> *And he answered them, saying, Who is my mother, or my brethren? And he looked round about on them which sat about him, and said, Behold my mother and my brethren! For whosoever shall do the will of God, the same is my brother, and my sister, and mother. (Mark 3:33-35)*

Jesus specifically stated that He considered those who do the will of God to be His family! How can you know if you are in His family or who else is in your family if you can't determine the will of God?

We have said the Bride is the ultimate community. Notice, every cell in your body is not in direct fellowship with every other cell. For example, a cell in your eye is not directly connected to a cell in your little toe. However, every cell in your body is in fellowship with every other cell through intermediary cells.

What do we call a group of cells in the body that aren't in fellowship with the rest of the body? Cancer. Ideally, you ought to be able to show the path of fellowship that connects you to another member of your body because you don't have to be in direct fellowship with every member of the body.

Once you have helped a person embrace truth and plug into community, leadership would teach this person doctrine that will help them get into God's will. The most powerful doctrine to teach them is the good, pleasing, and perfect will of God because it will show them they are in control of how much tension God needs to bring in the future in order for them to toss aside their plan and embrace God's plan for their life.

Now that you know this, you can state your will, whether you are an enabler or leader, through your actions for everyone to see, including the Spirit of Truth. In fact, taking all of this together results in my greatest revelation relative to God's will.

The Answer

The simplest way to bring about God's will in your life is two steps:

1. Give up control over the effects so the Holy Spirit can do His part (things you can't change).
2. Get as much spiritual value as possible (things you can change), which is the ultimate in addressing the causes!

The reality is, if you are focused on getting as much spiritual value as possible as a cause, you won't have time to try to take control from the Holy Spirit over the effects! I call the ultimate way of getting as much spiritual value as possible: Receive All Their Value.

When someone asks me for help (Step #7), I allow the Holy Spirit to give them truth through me. Notice, if they take the advice, then I have a share in their life. More specifically, I have a share in everything they accomplish that is an effect of getting over their issue. This is generativity and a way of creating spiritual value nearer the $100 bill level. Where it gets interesting is when they don't take my advice.

For example, if a person asks for my advice for how to help a group accomplish a project. I may explain to them the definition of a leader and how to communicate to avoid abuse. If they take this advice, I will have a share in everything this group accomplishes because of this advice. I will also have a share in everything the leader and the members accomplish apart from this project that relies on the definition of leadership and how to communicate to avoid abuse. What if they don't take my advice?

If I argue or get mad at them for not taking my advice, I have violated Step #6: Give up Control; Faith and Humility. Instead, I can state my will (Step #5) to forgive them (Step #3) resulting in avoiding a disagreement (Step #2). The easiest way to do this is to recognize the cause of the advice: the Holy Spirit! The person isn't ignoring me. They are ignoring the Holy Spirit! This was the same point God made to Samuel when the people said they wanted a king. They were rejecting God, not Samuel.

One thing I continue to learn each year is no one owes me the act of listening to me. My former belief in being owed in this area was unjust and an opportunity to lose spiritual value. If I trigger and get mad when someone doesn't listen to me when the advice coming through me is from the Holy Spirit, it's a missed opportunity for me to gain extraordinary amounts of spiritual value!

My understanding that the Holy Spirit can only move in response to justice (Step #1) results in me treating their ignoring of the Holy Spirit through me like a prayer (Step #4). I give up control by forgiving, and this is when it gets interesting.

If the person has another approach other than my advice and they can state why they are doing their approach and they actively do their approach, then they can remain under mercy. However, if they can't state why they are doing their approach, they are acting as if they are God.

Worse, if they don't do another approach or don't do the approach I offered *after* their approach has failed, then they are actively blocking the profitability of the advice that was given, which is an injustice, and the longer they go without doing the advice, the greater the injustice and the more spiritual value I receive through justice.

Even worse, if others come to them for advice and they don't pass on the advice I gave to them, they are blocking the profitability of the advice that was given by the Holy Spirit through me, which is a greater injustice, especially if something catastrophically unprofitable happens to this third party.

In my example, if the person doesn't take my advice, I avoid arguing with them and give up control over their actions. Whether they are a Christian or not, they will have to pay the difference between what could have been accomplished using my advice and what they actually end up accomplishing.

As much as that looks like a goldmine of spiritual value, there is a better way to receive all of a person's spiritual value. The person needs to be someone in authority who chooses to actively tell others not to do the advice you gave to them and also speak against you on a personal level for giving the advice! I would be due the difference between where everyone could have been if they followed grace through me and where everyone ended up without the advice. Would a Christian actively tell others not to take advice that came from the Holy Spirit?

When this happens, it is possible to receive *more* than all a person's spiritual value. Even though the person doesn't have enough spiritual value to cover their injustice, I can still receive all the spiritual value that I'm due through justice, just like Jesus did on the cross. This is similar to a person charging more on their credit card than what they have to cover the debt. Ultimately, this is the way to achieve the greatest spiritual value because someone is intentionally blocking generativity by being a boss!

As I began to understand this, it became exceedingly easier to give up control to the Holy Spirit and instantly forgive others. In fact, I spent that year changing all of my trigger responses to "forgive." It is what Jesus did while He was being killed!

Now Matthew 5:11-12 makes so much sense.

> *Blessed are ye, when men shall revile you, and persecute you, and shall say all manner of evil against you falsely, for my sake. Rejoice, and be exceeding glad: for great is your reward in heaven: for so persecuted they the prophets which were before you.*

Summary

1. Everything that is profitable in the long term happens by the Holy Spirit.
   - If you think everything happens by your effort, intelligence, and ability, you will wake up twenty years from now wondering how you wasted twenty years of your life.
   - Focus on getting spiritual value as a cause, and the Holy Spirit will take care of the effects better than any being in the universe.

2. My only goal every day is to receive everyone's spiritual value, including yours.
   - The people who don't let me receive their spiritual value are the ones I partner with to gain more value through profitability.

What do you think of the way I'm going about gaining spiritual value?

It turns out, God brings about His plan in the same way!

God gives the truth in love to everyone.

- Those who give back by allowing Him to flow through them in grace to bring about His plan are acting in truth and have directly facilitated God's will!
- God forgives those who intentionally choose to resist Him, and God accumulates spiritual value at their expense through justice so that He can bring about His plan.

> People who are opposing truth to their own detriment are indirectly facilitating God's will! Especially if they are in positions of authority!

All I'm doing is replicating how God brings about His plan.

I'm replicating God's will!

You can too, if you share God's doctrine in love and forgive people in authority when they reject God flowing through you! After all, would a believer *reject* God flowing through you?

Now that you understand the biggest system, you have the tools to explain anything!

Dr. Joel Swokowski's Commentary

Go Into All the World

Lenhart has given the ultimate approach and stated it represents the biggest system (God's will): Truth in Love and Forgiveness. Let's check his work!

Dr. Ackoff stated the first approach to any problem is absolve: ignore it and maybe it will go away. Notice, if you love, then absolve is not an option.

The second approach is resolve. This is when you focus on the effects, which is shown when you punish or reward. Basically, you are trying to equal out justice. Notice, if you forgive, then resolve is not an option.

The third approach is solve. This is when you look for the tangible cause and fix it. The problem is the solution only works for the tangible context.

Now, everyone involved has to deal with the consequences of the contextual answer in another context! This is why when you solve a person's problem, you create three more problems (Law of Unintended Consequences) and the overall stress from your solution is worse than from the original problem! Solve is not an option. We need an answer that is contextless!

We saw that truth looks for the intangible cause. Focusing on the intangible cause means the answer is contextless and the problem no longer appears in *any* context. It dissolves!

This "Truth in Love and Forgiveness" approach results in dissolve, which is the ultimate proof it is mada and therefore God's will!

The original core leaders of Music of Life Church (MOLC) have been taking territory for God. We have been vessels for God's will. The boundaries have expanded nationwide, into Canada, and even Europe. Through this expansion of the mission God has given us, we've learned a huge lesson in how God's will happens and in how to deal with the multitudes that reject us.

After learning the lesson that follows, I recorded a three-part video series and sent it to our MOLC Community. Here's that teaching that I named "Community Revelation":

Point #1 of this Community Revelation:
The belief that we have information that can change the world causes people to want to get this information out to others.

"What if people don't want it?"

First, I don't get to use that excuse if I'm not sharing the information (truth) myself. If you are using other leaders' experiences and using those as if they're yours, that's wrong. Please, tell their stories, but don't pretend they are your stories.

I have been rejected, many of the MOLC leaders have been rejected, and every time I tell those stories of rejection without supplementing it with my rejoicing, I am wrong!

I am struck by how bold the people of the UK are in sharing their stories and sharing the truth that has transformed their lives. In fact, it's causing me to wonder if I've been complacent.

Why is a community in the UK growing at the rate they are after just 6 months?

Why is that community growing faster than our local community after we have been living this information (truth) for over 15 years?

Am I showing God how much this information means to me?

Well, yes, I am…and my lack of amplifying the message He's given me is showing Him how little I care.

Question for you: How are you showing God that you appreciate the truth He's given you?

Point #2 of this Community Revelation:
Dealing with Rejection.

I've done an entire sermon on building a community. The three steps I presented were:
1. Meet a person to connect with them.
2. Get to know them to find out what they uniquely need.

3. Connect them with the person in the community who will best help them with that need.

Due to uniqueness, it's more likely someone else will be the person they ought to get help from, other than me. Not that I shouldn't be equipped to help everyone, but what I'm talking about here is who is *best* to help the person.

Am I a champion for others? Am I a promoter of others?

Or am I just wanting the credit myself?

I get it. I hate being rejected. I do want people to receive the truth and be transformed. But I feel like every time I try, they don't want it.

How do I deal with rejection?

Jesus was rejected. In fact, He needed to be rejected. I thank God He was rejected!! It led to the greatest value ever, literally ever, an infinite amount that paid for salvation for all!!

That's how I can deal with the rejection I experience after sharing truth: see it as a way I can experience a Christ-like life, that I'm living Jehovah's story…

…and see it as a measure that I'm sharing truth.

Everyone is qualified to share truth. You don't have to teach doctrine. You ought to share truth from your own story. Everyone is qualified to share at their own level of faith (understanding and experience).

Review Community Revelation
1. The belief that we have information that can change the world causes people to want to get this information out to others.
2. Community is built by connecting people into the community with someone who is best for them because they need to live out this information.

What if I get rejected?

Feeling great about being rejected would also be a measure of my speaking the truth in love. It's God's truth. He's the One being rejected!!
1. Speak Truth
2. In Love
3. Forgive

That's how I deal with rejection: Forgive… and start over with someone else!

Continuing to share proves I'm doing it for others' benefit, not for my own!

Continuing to promote others and champion others proves I'm sharing truth for the other person's benefit. This isn't about me. This is about expanding God's Kingdom!!!

Question for you: How have you dealt with rejection? Is it causing you to clam up and not share? Or are you simply seeing it as a speed bump, not stopping your progress, just a minor delay?

Question for you: When's the last time you championed or promoted someone else?

Point #3 of this Community Revelation:
Generating the Youthful Zeal.

Do you remember how it felt when you were first learning this truth? When you were finally getting in control of your life? Do you remember what it feels like when you're helping someone else successfully grow?

This is key to continuing to have that youthful zeal.

Having new people getting this information coming into the fold causes the community to retain this feeling that they are going to change the world!

Breaking this chain at any point causes the community to not grow and not feel like they can change the world.

Points 1 & 3 in this community revelation give the emotional aspect we need, not only to continue to grow our passion, but to feel great about the rejection we'll face when we share truth.

We've seen the way to handle rejection is to forgive, and then continue to speak the truth in love.

Point #3 is the key for us wanting to handle rejection the right way.

The key to us feeling great about the rejection we will face is a belief in the value being generated when we get rejected for sharing truth:

- share truth and people reject you: you get value for what could have happened with that truth.
- share truth with a leader and that value multiplies exponentially!

That value is what God can use to bring about the desires of your heart

That value is what God can use to make you happy

Humans are made to be truly happy when they are being who they are uniquely created to be. No one knows this better than God. Since God is right and just, He would achieve His will by working through a human in a way that accessed their uniqueness and resulted in joy for the individual. I think we called this dissolve!

(Or do I still think *my* plan is the best plan for my happiness?)

When rejected, in the short term, the truth is hindered; yet, the value I accrue can be used by God to get the truth out in other areas in a much greater way!

Community Revelation

1. The belief that we have information that can change the world causes people to want to get this information out to others.
2. Community is built by connecting people into the group with someone who is better for them because they need to live this information.
3. Having new people getting this information coming into the fold causes the community to retain this feeling that they are going to change the world.

Question for you: What are you going to do to grow a more positive emotion to the rejection you're bound to face?

You can build your faith through understanding, but eventually, like with self-esteem, you need to DO it... GET THE EXPERIENCE!!!... GIVE GOD THE VALUE AND SEE HIM MOVE...

Do this, and be ready for the natural to become supernatural.

Be ready for transformation in your own life.

Be ready for transformation in the lives of others.

The Next Book

THE NEXT BOOK is *Modeling God's Plan*. That book picks up where this book left off and it isn't that hard to see why. People who commit to generating spiritual value for God's will want to know what exactly they are working for. What specifically does God's plan look like? What does paradise look like? What does eternity look like?

It turns out God's plan is a kingdom made up of physical and spiritual beings. A lot of people who claim to be experts on the Bible are actually only aware of the physical realm as it relates to God's plan. That is only one-third of the plan. There is a spiritual plan and a plan where the physical and spiritual intersect!

For example, what do you think when you see evil increasing in the world? Most people feel like we are losing. Actually, we are winning!

In the final edition, we will see how the enemy has already made all their moves in the spiritual realm and is holding off making them in the physical realm for as long as possible. When we allow more of God to enter this physical realm, evil has to increase to combat it, which means we are getting closer to God's plan.

We have seen that God's plan is going to happen, so those who panic over what they believe is the enemy winning are showing people their lack of spiritual understanding and/or faith.

Evil beings know they are destined for eternal torture. How would you act if you knew you were going to jail for life? I think most people would try to put off the day as long as possible and enjoy themselves as much as possible. Likewise, the longer it takes evil to play their hand in the physical realm, the more people that will end up in the lake of fire.

Ultimately, it is our job to get as much spiritual value as possible so God can bring about His will, causing the enemy to play their physical cards quicker.

This series began with an analogy involving a wall. The goal of this book was to help you with the most uncomfortable portion of your journey away from the wall until you can see the entire picture. The final edition will take an analysis approach by stepping back towards the wall, filling in the missing tiles, and ending where this series began proving that we have explained the Christian worldview at the bottom rung.

www.ingramcontent.com/pod-product-compliance
Lightning Source LLC
Chambersburg PA
CBHW071354300426
44114CB00016B/2062